A HISTORY OF ELECTRICITY AND MAGNETISM

Burndy Library
Publication No. 27

A HISTORY OF ELECTRICITY AND MAGNETISM

Herbert W. Meyer

Foreword by Bern Dibner

BURNDY LIBRARY
Norwalk, Connecticut
1972

This book was designed
by The MIT Press Design Department.
It was set in IBM Composer Bodoni
by Science Press
printed on Mohawk Neotext Offset
by The Colonial Press Inc.
and bound by The Colonial Press Inc.
in the United States of America.

ISBN 0 262 13070 X (hardcover)
Library of Congress catalog card number:
70-137473

FOREWORD

Of the many ages of man—the Stone Age, the Bronze Age, the Iron Age, etc.—that preceded the 1800s, and that led one into the other, none was as rewarding to mankind as the electrical age. We now stand in awe of the space age, and in fear we face the nuclear age. From electricity, however, has been drawn an ever growing abundance of light, power, warmth, intelligence, and medical aid—all beneficent, silent, and ready.

Electricity is the one force in the arsenal of man that found no precedent in earlier history, nor was it drawn from classical times. It is fully the fruit of the Enlightenment, in time and place, and it generated its own enlightenment by extending man's waking hours and making him master of his own dawn and night. With no more than the touch of his finger he can summon energy almost without limit and can as readily cancel his summons. He can control his environment, be cooled when he wishes it or be warmed when needed. He can, in an instant, speak to anyone in any location where the fine filaments that carry this new force have been extended. He can, at will, observe and listen to public figures presenting problems of state or be entertained by whatever his choice of talent might be. No czar or emperor could command more.

With this new force man has probed the universe around him and has been compelled to change his estimates of its size exponentially. With radio astronomy he has penetrated distances measured in billions of light-years. He has probed the elemental nature of matter and energy; his genius challenged by their complexity, he devised new electronic probes and analytical instruments. His networks of computers have extended his intellectual powers (if not his wisdom) beyond all philosophic dreams. His weapons of destruction are diffused in hundreds of locations in the ground, in the sea, and in the air, each weapon's destructive strength measured in the equivalent megatons of explosive and all controlled by electrical energy equivalent to the power of a flea, the signal traveling over a hot line.

With such knowledge, such power, such control of environment, and such fleet means of communication, man has demonstrated his ability to reach the moon, travel its harsh surface, and return to earth—recording and televising his position, observations, and thoughts during the entire journey. This, and more, was realized in only a little more than a decade from the decision to attempt such a difficult mission.

Admittedly, the acquisition of electricity as an instrument of power and control in the inventory of man's abilities was no small addition. One can therefore stop and inquire about the circumstances that brought this acquisition about and review the events and personalities whose labors revealed the characteristics of a force unknown throughout the earlier millenia of work and study.

Two events ushered in the interest that blossomed into the arcane realm of electricity and magnetism. The first was the publication in London of a book on electricity and magnetism written by the physician to Queen Elizabeth who had devoted his leisure time and much of his fortune to investigating the properties of magnets and electrified bodies. Dr. William Gilbert's book, *De Magnete*, was published in Latin in 1600; its strength lay in the thoroughness with which the author examined each claim made in earlier writings on electrical phenomena, magnets, and compasses and in the exhaustive experimentation that separated the undisputed results of his observations from the accumulated superstitious claims of the past. This launched the science of electricity and magnetism from a firm basis of performance derived from objective experimentation.

The second event of significance that pushed electricity on its way toward usefulness was an announcement made in 1800 by Alessandro Volta that a new form of electricity could be drawn from a pile of alternating zinc and silver disks stacked one on the other, each pair separated from the adjoining pair by a cloth or paper disk saturated in brine. From the ends of this pile, Volta could draw a continuously flowing electric current, which he and others soon used to decompose water, to cause charcoal to glow with intense light in an electric arc, and, later, to deposit metals by electrolysis.

These two events–Gilbert's book on magnetism (1600) and Volta's constant-current electric cell (1800)–represent two centuries that neatly bracketed the nascent phase of electrical development. The one formulated more accurate knowledge about a force then useful for navigation in an era of voyaging and exploration; the other changed the concept of electric generation from frictional electricity, giving off bigger and bigger sparks, into an electric source of vast potential. However, the period must not be closed without tribute being paid to Franklin, an intrepid experimenter, who identified lightning as electricity and with his lightning rod helped rescue mankind from the terror of destroyed homes, steeples, and other structures. At the same time he helped guide man's faith to the truer character of natural forces and away from the ancient superstitions about lightning and the gods' intentions.

The next important step on the ramp of electrical progress was the discovery by Oersted in 1820 that a wire connecting the ends of a voltaic pile was enveloped by a magnetic field. It was then found that if such a wire were looped into a coil the magnetic strength was greatly intensified. With such a magnetic field Faraday showed, in 1831, that a moving electrical conductor had an electric current induced in it; the electric generator and its important adjunct, the transformer, were thus born. The generator supplied electric current in abundance and with this and auxiliary current from electric batteries the electric telegraph developed—the first important instantaneous disseminator of human intelligence over long distances. There followed the laying of the first transatlantic cable, the telephone, the electric lamp, and the electric motor. Each development generated a family of by-products—electroplating, the electric tramway, the moving picture—and each created a corresponding major industry.

As the drama of electrical development unfolds, we relate each forward step to the genius and perseverance of some experimenter, some inventor, some innovator. To him should go all the honor of a grateful people, for these are the true heroes of the modern world. The pages that follow will unfold their story, their struggles and attainments. May the telling never end.

Bern Dibner

PREFACE

PURPOSE OF BOOK

It is the purpose of this book to attempt to give to the reader, who has arrived upon the world scene in the midst of a scientific explosion, a sense of perspective and direction. The word explosion as it is used here is not entirely accurate because it implies a sudden, violent, and instantaneous event. The burgeoning of science resembles more nearly the propagation of a new grain from a few seeds, to a few bushels, and finally to a tremendous harvest.

A history such as this might be presented as a collection of biographies or as a series of stories concerning inventions and discoveries. It could discuss the unfolding of events from the standpoint of pure science or it might be weighted on the side of technology. All of these considerations have played a part in the writing of this story. Hopefully the book has blended these differing viewpoints in such a way as to stimulate the interest, not only of the student of science, but also that of the casual reader, who finds himself surrounded by the fruits of science and technology, knowing not from whence they came.

EXPERIMENTAL AND THEORETICAL RESEARCH

During the early period of the development of electrical and magnetic science, beginning with William Gilbert and Otto von Guericke, discoveries were the result of experiment coupled with observation and interpretation. At about the same time there began a different kind of scientific exploration based largely on mathematics, exemplified by Kepler's planetary laws and Newton's laws of motion. In electrical science mathematical analysis based on experiment came a little later in the work of such men as Ampère, Coulomb, Biot and Savart, Gauss, Weber, and Ohm.

Maxwell, an ardent admirer of Faraday's great genius, interpreted Faraday's discoveries mathematically and contributed his own mathematical findings, but credit for some of Maxwell's discoveries must be shared with Helmholtz, whose versatility in science has rarely been equaled. After Maxwell and Helmholtz followed a period of twenty or more years of fruitful experimentation. As a result of the Michelson-Morley experiment, Lorentz and Einstein brought forth new concepts of length and time that seriously upset the Newtonian system. Under the new theory length and time were no longer absolutes but were relative, and Newton's laws were valid only as special cases. Even more upsetting was the suggestion that matter and energy were convertible, one into the other.

These great triumphs of theoretical science were only a beginning, and there was much more to come. The nature of matter and energy became the primary object of physical research with most astonishing results. This type of research began with the work of J. J. Thomson, Planck, Lenard, Moseley, Rutherford, and Bohr, followed a few years later by the astonishing revelations of A. H. Compton, C. T. R. Wilson, de Broglie, Schrödinger, Heisenberg, Pauli, Born, Dirac, Davisson, Germer, G. P. Thomson, and Kikuchi.

MERGING OF THE PHYSICAL SCIENCES

As the nature of matter and its relationship to energy became clearer, the gaps between the branches of physical science began to close, and the lines of demarcation became indistinct. It is therefore difficult in a work such as this to avoid being drawn into byways only remotely connected with the subject matter.

ACKNOWLEDGMENTS

The author wishes to express his appreciation for the encouragement and suggestions given by Dr. R. J. Collins, head of the Department of Electrical Engineering of the University of Minnesota, and especially for important help given by Dr. W. F. Brown, professor of Electrical Engineering, also at the University of Minnesota. I also owe a debt of gratitude to my wife, Elfriede, for her valuable assistance.

1

Early Discoveries

ARCHAEOLOGY AND PALEONTOLOGY

Recorded political history now reaches back to about 4000 B.C., but we have some knowledge of mankind of much earlier periods based on the findings of archaeologists and paleontologists, aided by the studies of anthropologists. Prehistoric man fashioned weapons, tools, utensils, and clothing from stone, shells, bone, wood, skins, and sinews. Later he learned to use metals, such as copper and gold, which were found in their native state and required no smelting. Bronze, which is usually an alloy of copper and tin, came into use in Europe by the year 2000 B.C., or perhaps earlier, and gave rise to the period known as the Bronze Age.

There is no reliable record as to the date when the smelting of the ores of copper, tin, lead, and zinc began, but these metals were definitely in use before the beginning of the Christian Era. The Romans mined copper or copper ores in Cyprus and later obtained tin from Cornwall in England.

It is difficult to determine the date of the earliest iron implements, but iron artifacts have been found in Egypt dating back to 4000 B.C. Iron in its native state occurs only in meteorites and in tiny needles sometimes found in basalts. Little is known as to the date when man first produced iron from ores, but there is evidence that by 1350 B.C. the Hittites in Asia Minor succeeded in reducing the oxides of iron.

MAGNETITE AND THE LODESTONE

There are probably few natural materials known today that were not also known to prehistoric man, although he had little knowledge of their uses. Among the rather widespread and fairly abundant minerals of the earth is a very useful ore of iron called magnetite, which has the composition Fe_3O_4. It is a crystalline mineral, very dark in color, having a metallic luster, and a specific gravity equal to about five-sevenths of that of iron. Unlike any of the other iron ores it is magnetic, and this property gives it its name. There are occasional pieces of magnetite, as found in nature, which are permanently magnetized, and such pieces are known as loadstones or lodestones. In some specimens the magnetism is sufficiently strong to enable them to lift several times their weight of iron from one pole. There is good reason to believe that the attractive powers of lodestone had been observed as much as 3000 years before the Christian Era, but the directive properties of freely suspended pieces were probably not known until much later.

We are told that Huang-ti, or Hwang-ti, or Hoang-ti, an emperor of China in the year 2637 B.C., had a chariot upon which was mounted the figure of a woman, pivoted or suspended so that it was free to turn in any direction. The outstretched arm of the statue pointed always to the south under the influence of concealed lodestones.

Similar magnetic cars are mentioned by Chinese historians at various times down to the early centuries of the Christian Era. The Chinese also discovered that steel needles could be permanently magnetized by the lodestone and used as compasses. According to Humboldt, Tcheou-Koung, Chinese minister of state in 1110 B.C., used a compass employing a steel needle, and the same authority relates that Chinese mariners with the aid of compasses navigated the Indian Ocean, in the third century A.D.

Unlike the ancient civilizations of the West which perished, the Chinese civilization has continued in an almost unbroken line down-to the present time, so that its recorded history is far more complete than that of Babylon or Egypt. There is some disagreement among scholars as to whether or not the stories concerning the use of magnetic chariots or magnetic needles, by the Chinese, is fact or fiction. Since, however, such devices are possible, and since such stories recur from time to time in Chinese history and legend, there seems to be little reason to reject them.

THALES OF MILETUS

In 600 B.C. Greek civilization and commerce were flourishing. On the Grecian peninsula the city-states of Athens and Sparta had grown to greatness and power. Numerous Greek colonies had been planted on the shores of the Mediterranean and Aegean seas, among which was Ionia in Asia Minor. Miletus was a thriving seaport in Ionia, in and out of which sailed ships from all of the ports of the Mediterranean. Its inhabitants, through their trade with other countries, became well-to-do and acquired much of the culture of other civilizations of the Mediterranean basin. Philosophy, astronomy, mathematics, poetry, and art were cultivated along with commerce.

One of the inhabitants of Miletus in the year 600 B.C. was the philosopher Thales (640?–546 B.C.). A contemporary of Draco, Solon, and of King Nebuchadnezzar of Babylon, he was the founder of the Ionian school of philosophy from which Socrates came. Thales traveled extensively and received an important part of his education from the priests at Memphis and Thebes in Egypt. From them he learned a great deal about the physical sciences and geometry, and in the latter he soon excelled his teachers. Thales's life was devoted to teaching, discussion, philosophy, and statecraft. Unfortunately for posterity he left no writings, and all that we know of him was transmitted orally until Aristotle recorded his teachings.

Thales is important in electrical history because he was the first person who is said to have observed the electrical properties of amber. He noted that when amber was rubbed it acquired the ability to pick up light objects, such as straw, dry grass, and the like. He also experimented with the lodestone and knew of its power to attract iron. He apparently associated the two phenomena, although more than twenty-four hundred years elapsed before any actual relationship was proved. We do not know whether Thales discovered these facts for himself or learned of them from the Egyptian priests or from others. He apparently did not know of the directive power of the lodestone.

It is from the Greek that our terms electricity and magnetism are derived; the Greek word for amber is ελεκτρου (elektron), and the word magnet is thought to have come from Magnesia, a district in Thessaly, in which lodestones were found. According to Pliny, however, the word is derived from Magnes, the name of a shepherd who observed that the iron on the end of his crook was attracted by certain stones on Mount Ida that proved to be lodestones.

ANCIENT AND MEDIEVAL RECORDS AFTER THALES

The earliest written record that mentions the electrical properties of amber came from Theophrastus (about 300 B.C.). Later, various other Greek and Roman writers alluded both to the electrical properties of amber and the magnetic properties of the lodestone. Pliny the elder (23–79 A.D.), a Roman naturalist whose untimely death occurred on the seashore at Retina not far from Pompeii, in the eruption of Mount Vesuvius, referred to the attractive powers of amber several times in his *Natural History*. He wrote that the Etruscans, about 600 B.C., were able to draw lightning from the clouds and to turn it aside. We learn from others that the Temple of Solomon may have been protected from lightning by numerous sharp points of metal that covered the roof and that were connected by means of pipes to caverns in a hill. The temple of Juno is said to have been similarly protected. Lucretius, the poet and author of *De Rerum Natura*, noted the ability of the lodestone to attract several iron rings, one adhering to the other, and marveled at the peculiar behavior of iron filings in a brass bowl when a magnet was moved about beneath it.

As the great Roman Empire declined, the culture of Greece and Rome gradually vanished. Learning almost disappeared and for centuries was confined largely to the monks and priests in the Christian church. After the rise of Islam, in the early part of the seventh century, there was a great upsurge in learning among the Arab peoples, who, even though they destroyed the library at Alexandria, translated the works of the great pagan philosophers. Arab culture flourished during the Dark Ages when Western learning was at low ebb, where it remained until the beginning of the Renaissance.

There is an interesting item from Chinese history during the early part of our era. Koupho (295–324 A.D.), a distinguished Chinese naturalist, compared the attractive power of the magnet with the ability of excited amber to attract mustard seeds. From his manner of writing, it appears that this property of amber was no new discovery, but it is the first time the phenomenon was mentioned in Chinese history.

During the Middle Ages the properties of amber and the lodestone were not forgotten, but no new knowledge was added. St. Augustine in 426 A.D. expressed wonder at the ability of the lodestone to hold several iron rings suspended from it, and he mentioned an experiment in which a bit of iron laid on a silver plate is made to follow the movements of a magnet beneath the plate.

That long and dismal period of European history, generally called the Dark Ages, or the Middle Ages, came to an end sometime during the thirteenth or fourteenth centuries. In the world of science there may be some disagreement as to who was the first real scientist in the revival of learning, but surely Albertus Magnus (1206?-1280) was among the earliest. He taught at the University of Paris, where he was a distinguished professor of theology and philosophy and was also among the most learned in the science of the day. Also at the University of Paris at the same time was Roger Bacon (1214?–1294), who had first studied at Oxford. He became well versed in the scientific works of the Greek philosophers and the Arab scholars. It was Roger Bacon, more than anyone of his time, who insisted that human progress depended upon research and scientific education. His researches carried him into the fields of alchemy, medicine, optics, and mathematics. He is sometimes credited with the invention of the telescope and of gunpowder, although the former is generally ascribed to Lippershey of Middleburg, Holland and the latter to Berthold Schwartz.

In Roger Bacon's greatest work, his *Opus Majus* (1268), he reversed the manner of thinking of the Greek philosophers, which was largely subjective, to reasoning based on experiment. His writings and experiments, however, got him into trouble with the Church and he was accused of practicing black magic. He had become a Franciscan monk upon his return to England in 1250 but was soon enjoined by his order from writing or teaching and thereupon returned to Paris, where later the ban was lifted by a new pope.

THE MAGNETIC COMPASS

One of Roger Bacon's teachers was Petrus Peregrinus, or Pierre de Maricourt, who had carried on numerous experiments in magnetism. Petrus Peregrinus was not only a teacher but also a soldier attached to the engineer corps of the French army. During the year 1269, while he was with the armies of Charles of Anjou, which were besieging Lucera in southern Italy, he wrote a lengthy letter from his camp to a friend named Sigerus de Foucancourt, at his old home in Picardy. The letter described in detail his experiments with magnets, the construction of a floating compass, and also a pivoted compass employing a steel needle. This compass was provided with a card not unlike the mariner's compass of today. An excellent translation of Peregrinus's letter has been made by Professor Silvanus P. Thompson.

The Italian historian Flavius Blondus writes that Italian mariners, sailing out of the harbor at Amalfi, used a floating magnet as a compass before 1269. Its invention was attributed to a fictitious person named Flavio Gioia of Amalfi. There is an inconsistency, however, in this account because the alleged invention occurred in 1302. Blondus asserts that the real origin of the magnetic compass is unknown. There seems to be little doubt that the Italians learned about the compass from the Arabs.

The magnetic compass was the first device having practical value that came from experiments with magnetism. After the time of Petrus Peregrinus the compass soon came into general use and led to many theories as to the reasons for its behavior. Its variations in different longitudes were noted, together with other changes of short duration. Undoubtedly the voyages of Columbus and Vasco da Gama were greatly aided by its use.

With the invention or rediscovery of the compass came a greatly increased interest in magnetism. Many believed that magnetism was a cure for various diseases, and others claimed to be able to use compass needles, separated by great distances, as a means of telegraphic communication. Still others, among whom was Petrus Peregrinus, claimed to have made perpetual motion machines using magnets. Robert Norman of Wapping, England, a maker of compass needles, was the first to make a dipping needle. He found the inclination of the needle at London to be 71 degrees, 50 minutes.

During the Renaissance epoch human progress was more rapid than at any earlier period. It was far more than a rebirth of the classical learning of Greece and Rome. Not only did art and literature flourish, but great new lands were discovered; printing, gunpowder, and the telescope were invented; and Copernicus gave the world a new concept of the universe. The Church was shaken, and autocratic government began to lose its despotic power. There was, however, little further progress in the science of electricity and magnetism until about the year 1600.

WILLIAM GILBERT

William Gilbert was born at Colchester in England in 1544. He studied medicine at St. John's College, Cambridge. After graduation he traveled about Europe for a time and returned to London in 1573, where he practiced medicine very successfully. At the same time he carried forward a series of experiments in electricity and magnetism and also studied all of the available writings of others on the subject. Gilbert devoted seventeen years of his life to compiling the results of his researches into a Latin volume entitled *De Magnete, Magneticisque Corporibus, et de Magno Magnete*

Figure 1.1 William Gilbert Demonstrates Electrostatic Attraction at the Court of Queen Elizabeth *(Courtesy Burndy Library)*

Tellure; Physiologia Nova, Pluribus et Argumentes et Experimentis Demonstrata. This formidable title is now generally condensed into *De Magnete*. It was published in 1600 and represented the greatest forward step in the study of electricity and magnetism up to that time. Gilbert regarded the earth as a huge magnet and explained the behavior of the compass on this basis. From his experiments in magnetism Gilbert deduced many ideas concerning the magnetic field, magnetic induction, polarity, and the effects of temperature on magnets. In his electrical researches, he found a long list of materials that could be electrified, which he called electrics. He devised a form of electroscope that he called a versorium, which was a pivoted nonmagnetic needle. He was the first to use the term electricity.

Gilbert's skill as a physician and his genius as a scientist were recognized by Queen Elizabeth, who appointed him as her court physician in 1601. He died in 1603.

Gilbert has been honored by the use of his name as the unit of magnetomotive force. In his *De Magnete*, Gilbert reiterated the plea of Roger Bacon for more intensive research. By a queer coincidence another Bacon, this time Sir Francis Bacon, who followed soon after Gilbert, published in 1620 a great work of science entitled *Novum Organum*. As far as we know, Sir Francis Bacon never carried on any original scientific research, but he set forth the scientific achievements up to his time so clearly and presented the case for research so eloquently that the book was, and still is, a source of inspiration for scientists.

2

Electrical Machines and Experiments with Static Electricity

Scientific progress during the later years of the sixteenth and throughout the seventeenth century was astonishing. Among the great names of science of that period were Copernicus, Gilbert, Brahe, Napier, Francis Bacon, Galileo, Kepler, Descartes, von Guericke. Torricelli, Boyle, Huygens, Mariotte, Newton, and Leibniz. During this period also, the first experiments on the steam engine were made by de Caus, Papin, and Savery. These experiments were carried to successful conclusions in the eighteenth century by Newcomen and Watt.

It is only fitting that in this recital of great men of science, special mention be made of Sir Isaac Newton, who is regarded by many as the greatest of scientists. His contributions to human knowledge were largely in the fields of mathematics, optics, astronomy, and in the laws of mechanics and gravitation. He made some minor experiments in electricity also, but his importance in this history lies principally in the effect of his work on research. His *Principia* was the third great scientific work published in England in the seventeenth century.

OTTO VON GUERICKE

In electrical science there was again a considerable lapse of time after Gilbert's *De Magnete* until Otto von Guericke, the burgomaster of Magdeburg, constructed the first electrical machine in 1660. Von Guericke is also distinguished as the inventor of the air pump and as the one who devised the spectacular Magdeburg hemispheres experiment.

Figure 2.1 Von Guericke's Electrical Machine *(From Burndy Library)*

Von Guericke's electrical machine marked the most substantial advance yet made in electrical knowledge. In this machine a sulfur ball that had been cast in a glass globe was mounted on a shaft which passed through its center. The ball was rotated by means of a crank at the end of the shaft. In later models the shaft was driven at higher speed by means of a belt that passed over a larger driving wheel and over a smaller pulley on the shaft carrying the sulfur ball. The rotating ball was excited by friction through the application of the dry hands or a cloth.

This machine produced far greater quantities of electricity than had hitherto been available and made possible new and interesting experiments. Von Guericke noted the attraction and repulsion of feathers, the crackling noises and sparks, and the odor that permeated the air when the machine was excited. He found also that the electrification of the ball produced a tingling sensation when any part of the body approached it. There is reason to believe that von Guericke noted that electricity from his machine could be transmitted several feet over a piece of string. These experiments were witnessed by many persons with lively interest, and news of the device soon spread to all parts of Europe.

Within a short time similar machines with variations and improvements were constructed by others. Sir Isaac Newton became interested in the experiments and is credited with the construction of an electrical machine having a glass globe about the year 1675.

OTHER EXPERIMENTS WITH STATIC ELECTRICITY

Jean Picard, a French astronomer, noted in 1675 that when a Torricellian barometer was agitated in the dark, flashes of light appeared in the evacuated space above the mercury. The mercury barometer was invented by Evangelista Torricelli of Italy in 1643. It consisted of a vertical glass tube closed at one end and filled with mercury. When the mercury-filled tube was inverted with the open end below the surface of the mercury in a cup, the level of the mercury in the tube dropped to a point at which it was sustained by atmospheric pressure, leaving a vacuum in the tube above the mercury. Francis Hawksbee pondered this phenomenon and in 1705 carried on a series of experiments to determine the cause of this light. He used glass vessels containing mercury, some of which had been exhausted while others had not. When the vessels were shaken, the light appeared in the evacuated vessels but only faintly in those containing air. The appearance of the light was also different in the exhausted vessels from that in the vessels containing air. In the vacuum the light was in the nature of a glow that permeated the evacuated space, whereas in the vessels containing air the light appeared as weak flashes.

Hawksbee discovered that similar effects could be produced on glass vessels without mercury simply by rubbing the exterior surfaces, and thereby proved the electrical nature of the phenomenon. He also showed that an evacuated glass globe could be made to glow by bringing it near another globe that had been electrified by rubbing. He performed many beautiful experiments showing colors and striations with varying degrees of evacuation and various shapes of glass vessels. He may have noted the similarity between these effects and the aurora borealis. Following these experiments, Hawksbee proceeded to build electrical machines using revolving glass globes. Some of his machines were powerful and produced sparks of considerable intensity.

Among other experimenters was Professor Johann Heinrich Winckler of the University of Leipzig who, about the year 1733, substituted a fixed cushion for the hand or cloths which had previously been used as rubbers. Georg Matthias Boze (1710–1761) of Wittenberg about 1745 added a prime conductor, with which greater quantities of electricity could be collected. At Erfurt a Scottish monk named Gordon constructed a machine using a glass cylinder rather than the previously common glass globe.

About the year 1670, Robert Boyle, whose chief fame rests on his work with gases, made some additions to Gilbert's list of substances that could be electrified and also found that the attractions between electrified and nonelectrified bodies were mutual. In the *Philosophical Transactions* in 1708, Dr. William Wall published his observations on the sparks and crackling noises emanating from electrified bodies, which he compared with lightning and thunder. Winckler made similar observations somewhat later, and suggested the use of conductors for protection against lightning.

STEPHEN GRAY AND THE TRANSMISSION OF ELECTRICITY

Stephen Gray (1695-1736), a pensioner at Charter House in London, some time prior to 1728 began a series of electrical experiments with very limited equipment. His earlier experiments were of a minor nature, which included the discovery of the electrification by warming and rubbing of such materials as feathers, hair, silk, linen, wool, cloth, paper, leather, wood, parchment, and goldbeaters skin. He found also that linen and paper could be made to give off light in the dark.

His most notable discovery, however, was that electricity could be transmitted. In 1729 he had made many fruitless efforts to electrify metals by the same methods he had used on other materials when it occurred to him that perhaps he could transfer a charge from an electrified glass tube to a piece of metal. He used for this purpose a glass tube 1$^1/_5$ inches in diameter and 3 feet, 5 inches long. A cork had been fitted into each end of the tube to keep out the dust. He tried first to determine whether or not there was any appreciable difference in the electrification of the tube with and without the corks and found none. He did find, however, that when the tube was electrified the cork would attract and repel a feather.

In his next experiment he attached an ivory ball to the end of a fir stick about four inches long and inserted the other end of the stick in the cork. When the tube was rubbed, he found that the ball was electrified as the cork had been. Carrying the experiment further, he attached the ball to longer sticks and to brass and iron rods, with similar results.

As the sticks and metal rods became longer, he experienced difficulty due to bending, and he conceived the idea of using a piece of packthread or string attached to the cork and to the ivory ball. With the longest packthread he could manage by suspending the ball over the edge of a balcony he was still able to transmit electrical charges to the ball. He then tried suspending a longer packthread over a nail in a beam to the ivory ball, but this time the experiment failed. He surmised correctly that the charge had been led off through the nail into the beam.

On June 30, 1729, Gray visited Granville Wheeler in the country to demonstrate his experiments. Together they worked to transmit the electrical charges over the greatest possible distance. Mr. Wheeler suggested suspending the packthread on silk threads, and this arrangement worked admirably. They had transmitted the charge over 80 feet of packthread. In order to increase the distance still further, they looped the packthread back through the same gallery, a total distance of 147 feet. In further experiments they finally reached a distance of 765 feet. In still other experiments they discovered that hair, rosin, and glass made suitable supports for their packthread line. On the same day, which was July 2, 1729, Gray and Wheeler electrified larger surfaces such as a map and a tablecloth.

In August of the same year, Gray found that he could produce charges at the end of an insulated packthread line merely by bringing the electrified glass tube near the other end of the line without touching it and that the electrification was greatest at the far end of the line.

Gray made another notable discovery in which he found that an iron rod, pointed at both ends and suspended from silk lines, gave off cones of light in the dark when it was approached by an electrified glass tube. He also commented on the similarity between electrical sparks with the noises they produced and flashes of lightning with peals of thunder. No better example of the value of research and observation may be found than in Gray's experiments with simple apparatus and an inquiring mind.

Gray made the important discovery in 1729 that some substances were conductors and others were nonconductors. He may have been the first to use wires as conductors. The art of wiredrawing was not discovered until the fourteenth century and did not reach England until the seventeenth century, so that in the time of Gray, wire was still a comparatively new item.

DU FAY'S EXPERIMENTS AND HIS DISCOVERY OF TWO KINDS OF ELECTRICITY

In Paris Charles Du Fay (1698–1739), a retired military officer and a member of the French Academy of Sciences, reported the results of his experiments to the Academy during the years 1733 and 1734. He disproved the statement by Jean Desaguliers that all bodies could be classified as electrics or nonelectrics by showing that all bodies could be electrified. In the case of conductors it was necessary that they be insulated. He showed that a string was a better conductor when it was wet and succeeded in conducting electricity over such a line a distance of 1256 feet. Among his other interesting experiments was the electrification of the human body when it was insulated from the ground. He noted that when another person approached the one who was electrified, he experienced a prickling sensation, and in a dark room there was an emission of sparks. Du Fay rediscovered an effect which had been observed by von Guericke, namely, that a charged body attracts another body, which after contact receives a similar charge and is then repelled.

Du Fay's most important discovery, however, was that there were two kinds of electricity, which he called vitreous and resinous. The first, he said, was produced on glass, rock crystal, precious stones, hair of animals, and wool. The second was produced on amber, copal, gum-lac, thread, and paper. He announced also that these electricities repel similar charges and attract opposite kinds.

IMPROVEMENTS IN ELECTRICAL MACHINES

In 1746 Dr. Ingenhousz made a glass plate machine, but the same invention has also been attributed to Jesse Ramsden, although this was not until 1768. Benjamin Wilson about 1746 invented a collector for an electrical machine somewhat resembling a comb. It consisted of a metallic rod with a number of fine points, so mounted that the points were close to the revolving electrified surface.

THE LEYDEN JAR

A very important discovery was made on November 4, 1745 by E. G. von Kleist of Kammin, Pomerania, Germany, which was first credited to Professor Pieter van Musschenbroek and his assistant Cunaeus of Leyden, Holland, but priority belongs to von Kleist. Nevertheless, the invention has since been known as the Leyden jar. It was found that a bottle partly filled with water and containing a metal rod which projected through the neck would, when held in the hand and the rod presented to an electrical machine, receive a powerful charge. So great was the charge that after the bottle was removed from the electrical machine the person holding the bottle would receive a severe shock when a finger of the free hand touched the central rod.

The news of this discovery spread so rapidly through Europe that within a very short time it had been repeated everywhere, and some individuals traveled throughout the continent demonstrating the new discovery, and gaining a good livelihood thereby. In England Sir William Watson showed the Leyden jar to Dr. Bevis, a colleague, who suggested coating the outside with sheet lead or tin foil to replace the human hand. Another experimenter, John Smeaton, applied tin foil to both sides of a pane of glass and obtained excellent results, whereupon Watson coated both the inside and the outside of several large glass jars with leaf silver and succeeded in storing powerful charges.

THE SPEED OF ELECTRICITY

The Leyden jar provided a new and useful tool for carrying on electrical research, especially experiments in the transmission of electricity. By this time the use of wires as conductors was commonplace. In France electricity from charged Leyden jars was transmitted a distance of 2½ miles. Pierre Charles Lemmonier, the French astronomer, attempted to measure the velocity of electricity but found that the time required to travel a distance of 5700 feet was inappreciable. In England Sir William Watson, together with Henry Cavendish, Dr. Bevis, and others, conducted similar experiments using the ground as one side of the circuit. Baked or dried sticks were used as supports for the wires. On August 5, 1748, at Shooters Hill, these men set up a circuit of 12,276 feet through which they discharged a Leyden jar and decided that transmission was instantaneous.

SIR WILLIAM WATSON'S THEORIES

In another experiment, Watson observed that bodies unequally charged with the same kind of electricity tend to equalize their charges when they are joined. Watson was also the first to apply the terms plus and minus to electrical polarities; therefore, he may have shared Franklin's view that there was, in fact, only one kind of electricity and that there appeared to be two kinds due to a relative excess or deficiency. Watson was the author of several books on electricity, one of which, published in 1746, entitled *Nature and Properties of Electricity*, first aroused Franklin's interest in the subject.

MISCELLANEOUS DISCOVERIES

Probably the first attempt to use electricity for telegraphic purposes over long circuits was made by Johann Heinrich Winckler of the University of Leipzig in 1746. In some of his experiments he used the river Pleisse for a portion of the return circuit. Following the work of Gray and Du Fay on the transmission of electricity, the idea of using electricity for telegraphic purposes occurred to a number of men. Before the discovery of voltaic electricity and electromagnetism the kinds of signals were limited in number. Winckler probably used sparks.

Pierre Lemonnier of France discovered that the quantity of electricity communicated to a body is not in proportion to its volume but in proportion to its surface. He also discovered that the shape of a body influenced its ability to receive a charge.

The Abbé Nollet (Jean Antoine 1700–1770), who was the friend and co-worker of Du Fay, made numerous discoveries and observations. He found that when an uninsulated body was introduced into an electrical field of influence it became electrified. This effect had also been noted by Gray. Nollet observed that when sharp points were brought into such a field, the sharpest were first to give off brushes of light. He may also have been the inventor of an electroscope consisting of two threads attached to a conductor, which diverged when electrified.

The Abbé also performed numerous experiments on the influence of electricity on the flow of liquids from capillary tubes. Nollet called attention to the similarity between electricity and lightning, as Dr. Wall, Stephen Gray, Winckler, and others had done.

There had been little interest in magnetism since the time of Gilbert, but there was one new development. Knight and Michell in England, and Duhamel in France, during the years 1745 to 1750, had constructed several powerful steel magnets by new processes of heat treatment and by the use of a number of smaller bars to build up a single larger magnet. The principal source of magnetism, however, was still the lodestone, except that some rather weak magnets had been made by striking steel bars held in the magnetic meridian.

The discovery had been made about this time that a current of air issued from an electrified point at the same time that such a point gave forth a brush discharge. Hamilton of Dublin used this discovery to construct the first electrically operated motor, which consisted of a wire stuck through a cork with the pointed ends bent in opposite directions. The axis was a needle that was suspended from a magnet so that it could turn almost without friction. When the points were electrified the device rotated so long as the electrification continued.

Benjamin Wilson made a somewhat similar device, except that he provided vanes on the cork set in motion by a stream of air issuing from an electrified point placed just outside them.

Hawksbee, Nollet, and others had observed the similarity between electrical discharges in a vacuum and the aurora borealis. Perh Vilhelm Margentin, secretary of the Swedish Academy of Sciences, addressed a letter to the Royal Society, dated February 21, 1750, in which he mentions his observations concerning the effect of the aurora borealis on the magnetic needle. If, as some believed, the aurora borealis was an electrical phenomenon, this discovery by Margentin established a relationship between electricity and magnetism. The discovery had also been made that pieces of steel that had been struck by lightning were sometimes magnetized.

BENJAMIN FRANKLIN'S EXPERIMENTS

Benjamin Franklin (1706–1790) of Philadelphia was the first American who made a notable contribution to electrical science. His fame in science rests chiefly on his demonstration that lightning was an electrical discharge, but this discovery had been anticipated by others, both in theory and in experiment. His well-known kite experiment was performed in June 1752, during a thunderstorm in a field at the outskirts of Philadelphia. He succeeded in charging a Leyden jar from the kite string as it began to rain and a heavy thundercloud passed over.

Franklin had written at some length concerning his theories regarding lightning and his proposed methods of proving them. These writings were published abroad before he was able to perform his experiment, with the result that others were soon engaged in similar endeavors. In France, Dalibard and Delor proved the electrical nature of lightning

Figure 2.2 Benjamin Franklin *(From Smithsonian Institution)*

about a month before Franklin's kite experiment, by means of elevated rods from which Leyden jars were charged. Canton and Wilson, in England, carried on similar experiments with good results, and in St. Petersburg, Professor Richmann was killed by lightning while experimenting with an elevated rod during a thunderstorm.

Other experiments performed by Franklin were probably of greater lasting importance. He began his work in electricity in 1746, at which time he had purchased some simple pieces of equipment after reading a book on the subject by William Watson. Perhaps his earliest discovery was the fact that when a piece of glass was rubbed with a cloth, the glass received a charge of the same strength as the charge on the cloth but opposite in kind. He concluded that all bodies contain electricity, and that when two dissimilar substances are rubbed together the one receives an excess of electricity and the other a deficiency. From this supposition he reasoned further that there was but one kind of electricity rather than two, as Du Fay and others had supposed. Dr. Watson in England may have come to the same conclusion when he spoke of plus and minus electricity, terms that were very similar to Franklin's positive and negative.

Franklin demonstrated his theory concerning a single kind of electricity by having two people stand on insulated platforms, one of whom rubbed a glass tube with a cloth and took the charge from the cloth, while the other took the charge from the glass. When they brought their fingers close together a strong spark passed between them and both were completely discharged, showing that the charges had neutralized each other. Another demonstration of the same principle was made by hanging a cork ball between two knobs connected to the inner and outer linings of a Leyden jar. The ball would vibrate between the two knobs until the Leyden jar was completely discharged and both coatings were neutral with respect to the earth.

When a Leyden jar is charged, the outer coating must be connected to ground or to some other equivalent body. Franklin showed that the inner coating will receive only as much electricity as is driven from the outer one. He noted that a linen thread, suspended near the outer coating, was unaffected until the equilibrium was disturbed by bringing a finger near the knob connected to the inner coating.

Franklin showed also that, apparently, the charge held by a Leyden jar was on the glass and not in the coatings. He demonstrated this theory with a jar in which water formed the inner coating. When the water was poured off, he found that the water was not charged. Upon filling the jar with fresh water he was able to discharge it. Franklin performed similar experiments using glass plates with removable coatings. Later research has shown that although the facts of the experiment were correct, the interpretation is slightly different. It is now known that the charge is of the nature of a strain in the dielectric.

In 1748, Franklin described an electrostatic jack, or motor. He attached two metal knobs to diametrically opposite sides of an insulating wheel and placed the wheel so that it would rotate between two stationary insulated knobs, with a very small clearance between the rotating and stationary ones. When the stationary knobs were electrified, one positively and the other negatively, the wheel was caused to rotate. The knobs on the wheel were first attracted, and as they approached the stationary knobs a spark passed, making the charge on both of the same sign. The result was that they repelled one another and caused the wheel to rotate.

ATMOSPHERIC ELECTRICITY

On April 12, 1753, Franklin succeeded in causing an alarm bell to ring by means of atmospheric electricity. He had installed an insulated iron rod at his house, which projected some distance into the air. The rod was connected to a device in which there were two bells, with a lightly hung, insulated clapper between them. One of the two bells was connected to the elevated rod and the other to ground. When the rod became sufficiently electrified, the clapper was attracted first to one bell and then to the other.

The elevated rod served for various other experiments involving atmospheric electricity. He found that the charge on a cloud might be either positive or negative and that a cloud could change the polarity of its charge. Franklin observed strong electrification of the atmosphere during a snowstorm and even when the sky was entirely clear.

In another series of experiments, Franklin became interested in the nature of the electric field surrounding a charged body. In order to visualize the shape of the field, he suspended electrified bodies in still air, and then filled the air with smoke from rosin, which he had dropped on hot plates. The smoke, much to his amazement and gratification, formed beautiful patterns about the charged body.

Franklin had a friend at Boston, a Mr. Kinnersley, who was also interested in electricity. They corresponded frequently concerning their experiments. Kinnersley rediscovered Du Fay's findings concerning two kinds of electricity by the use of two electrical machines, one with a sulfur ball, and the other with a glass globe. Neither of the two men knew of Du Fay's work. Franklin repeated Kinnersley's experiment, and observed that the sparks from the glass and sulfur balls were different in appearance. The brush discharge from the prime conductor of the glass globe machine was large, long, and divergent, and made a snapping noise, while that from the sulfur machine was short, small, and made a hissing noise. Kinnersley also experimented with the thermal effects of electricity, and by using a battery of Leyden jars he succeeded in melting fine iron wires.

For his many achievements in electrical science, Franklin was awarded the Copley Medal, England's greatest scientific award, by the Royal Society of London in 1753.

EXPERIMENTS IN EUROPE WITH ATMOSPHERIC ELECTRICITY

Lemonnier, the Abbé Mazeas, and Beccaria, during the years 1752 and 1753, performed many experiments on atmospheric electricity using both kites and elevated rods. They found evidences of electrification at most times, but especially when there were thunderclouds in the sky. Usually there was little electrification at night, but it increased greatly after sunrise and diminished after sunset.

Canton discovered that the air in a room could be electrified, as did Beccaria also. Canton showed that the air in a room acquired the same kind of electricity as a charged body in that room and that the divergence of two threads attached to the charged body gradually decreased as the air became electrified, even though the charge on the body was maintained by an electrical machine. Beccaria produced sparks under water and noted the formation of bubbles accompanying the sparks but did not suspect that the bubbles were due to the decomposition of the water.

ELECTRICAL INDUCTION

Electrical induction is the phenomenon in which a body becomes electrified when it is approached by another body carrying an electrical charge. Its effects had been noted by Hawksbee, Stephen Gray, Nollet, and probably by many others. It was now under investigation by Franklin, Wilke, Canton, and Aepinus. Experiments had shown that when a charged body was brought near an insulated conductor, a charge similar in kind to that of the charged body appeared at the remote end of the insulated conductor (repelled charge), and that if this charge were led off to ground and the original charged body removed, there remained on the insulated conductor a charge (bound charge) of sign opposite to the original one. Stephen Gray had in part discovered similar effects with his packthread line but had not observed the relationship of kinds, nor that there were two distinct induced charges. The discovery of inductive effects led to the invention of the electrophorus in 1775 by Alessandro Volta. It consisted of a flat metallic disk, or shallow pan, which was covered with sealing wax or rosin and provided with a removable disk of metal, of slightly smaller size, having an insulating handle at the center. When the sealing wax or rosin was electrified by rubbing with a catskin and the removable disk was brought down near the surface of the charged electrophorus, induced charges were produced in the disk. By touching the finger to the removable disk before it was lifted, the repelled charge was led off, after which the finger was removed and the disk lifted, which then had on it a free charge. The electrophorus was a useful instrument, because it provided rather large quantities of electricity with little effort.

ELECTROSCOPES

The Reverend Abraham Bennet described in the *Philosophical Transactions* for 1787 two inventions of great importance. One was the gold leaf electroscope that, with subse-

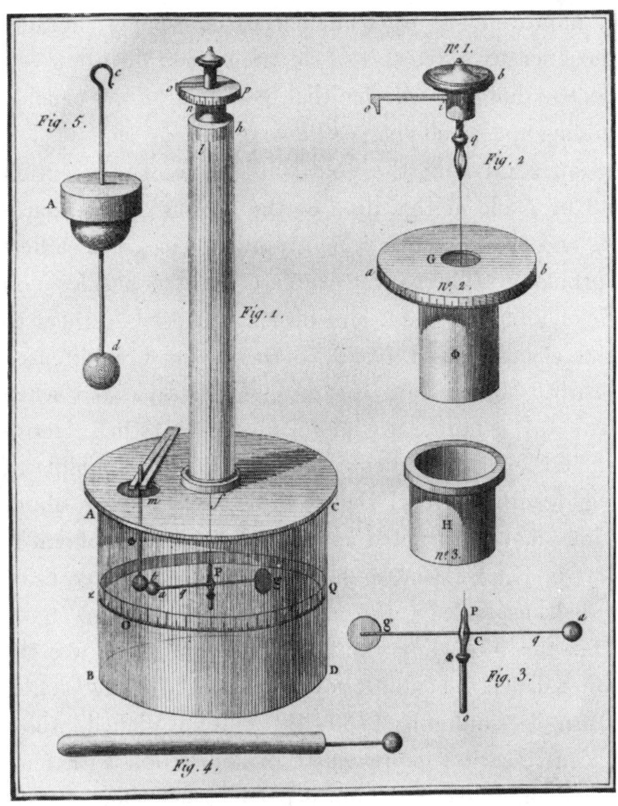

Figure 2.3 Coulomb's Torsion Balance *(Courtesy Burndy Library)*

quent improvements by William Hasledine Pepys, became the most sensitive detector of electricity, and the other was his electric doubler, a device that by induction was capable of building up a small charge to a large one.

In connection with the gold leaf electroscope, mention should be made at this time of the various other electroscopes and electrometers which came into use. The earliest was probably Gilbert's versorium, a pivoted needle. Von Guericke, Stephen Gray, and others had used feathers to indicate electrification. Nollet, Beccaria, and Franklin used two threads hung close together, which separated when electrified. It is not clear who invented the pith ball electroscope, but such instruments were in use by the middle of the eighteenth century. Daniel Gralath, of Danzig, about this time, had constructed an electrical balance. William T. Henley, in 1772, invented the quadrant electrometer, using a pith ball suspended at the center of a graduated arc.

By far the most important device of this kind was the torsion balance, invented about the year 1784 by Charles Augustin de Coulomb (1736-1807). John Michell, whose name was mentioned previously in connection with magnets, had described a torsion balance earlier which Cavendish used to determine the mean density of the earth. Coulomb used the instrument to prove the inverse squares laws governing electric and magnetic fields. This law had been announced for magnetic fields by Johann Tobias Mayer at Göttingen in 1760, and for electrical fields by Cavendish in 1762, but in each case with only partial proof.

Coulomb demonstrated with great accuracy that the force between two magnetic poles is proportional to the product of the pole strengths and inversely proportional to the square of the distance between them, and that the force between two electrical charges is proportional to their product and inversely proportional to the square of the distance between them. These laws are expressed in later terms by the equations $f = mm'/(\mu r^2)$ and $f = qq'/(Kr^2)$, in which f is the force, m and m' are the magnetic pole strengths, μ (the Greek letter mu) is the permeability of the intervening medium, r is the distance between the magnetic poles or electrical charges, q and q' are the electrical charges, and K is the dielectric constant.

Coulomb's use of the torsion balance marked the beginning of quantitative work in electricity and magnetism. Coulomb also devised a magnetometer for measuring the intensity of the earth's magnetic field. Charles Borda, a French astronomer, had made certain magnetic measurements with some success in 1776, and still earlier George Graham had suggested methods of measurement but had not carried them out.

OTHER DISCOVERIES IN THE EIGHTEENTH CENTURY

During the closing years of the eighteenth century there were many discoveries of varying degrees of importance. Georg Christoph Lichtenberg, professor of experimental philosophy at Göttingen noted in 1777 that when fine powders, such as lycopodium spores, were electrified and dusted upon surfaces electrified with the opposite kind of electricity, the particles arranged themselves in beautiful configurations. The positively charged powders showed figures resembling feathers, while those negatively electrified arranged themselves in starlike figures. These are known as Lichtenberg figures and are at present produced more clearly on photographic plates.

In 1781 Lavoisier, the great French chemist, showed that electrification was produced when solids or liquids were converted into gases or when gases were released during effervescence caused by chemical action.

Martin Van Marum, of Holland, about 1785 constructed the most powerful frictional electrical machine built up to that time. It was a plate machine, having two circular glass plates 65 inches in diameter, each of which was excited by four rubbers. So powerful was this machine that a pointed conductor at a distance of 28 feet showed a brush discharge. Van Marum also constructed a battery of Leyden jars with a total surface of 225 square feet. The discharge from this battery magnetized steel bars $^1/_{12}$ inch by 1 inch in cross section and 9 inches long.

The magnetic properties of cobalt and the diamagnetic properties of bismuth and antimony were discovered in 1778 by Sebald Justin Brugmans of Holland. A diamagnetic substance is one that has a permeability in air or in a vacuum of less than one. A bar or needle of such a substance when free to move would tend to take a position at right angles to the lines of force in a magnetic field.

Before concluding the portion of this history preceding the epochal discovery of the voltaic cell, mention must be made of Joseph Priestley (1733–1804). His fame rests chiefly on his discovery of oxygen, but he carried on a considerable amount of electrical research and wrote the first electrical history entitled *The History and Present State of Electricity*. This book of 736 pages was published in 1767, and notwithstanding the fact that electrical science had scarcely been born at this time the book is still well worth reading.

Priestley made only minor contributions to electrical science, among them the fact that carbon was a conductor, yet he showed a remarkable insight into the nature of electrical phenomena and predicted that new developments in the art would far outstrip anything that had so far been conceived. He could hardly have dreamed of the magnificent fulfillment of his prophecy.

During the period whose history has been traced in this chapter, electricity had advanced from the position of a curious and mysterious phenomenon, about which very little was known, to the foremost position in scientific interest. The long list of experimenters, already mentioned, had transformed electrical knowledge from a natural curiosity into a science. Notwithstanding the great progress that had been achieved, there was not yet a single practical application of electricity, unless the lightning rod, Franklin's electrostatic bells, or the whirligigs of Hamilton, Wilson, and Franklin could be considered as such.

3

Voltaic Electricity, Electrochemistry, and Electromagnetism

The closing decades of the eighteenth century were years of unusual importance to electrical science. Up to that time all known electrical phenomena originated solely from electrical charges produced by friction, heat, or induction, and currents of electricity were of a transitory nature, resulting from the discharges of accumulated electrical charges.

In the period now under consideration, new contrivances were discovered which were capable of producing, by chemical means, steady currents of electricity. The quantities of electricity available from the various electrical machines previously in use were minute, and therefore the chemical, heating, and magnetic effects were difficult to observe, although all of them, as a matter of fact, had been noted.

On numerous occasions, important electrical discoveries had been made without recognition of their value, and only after one or more rediscoveries of such phenomena did they lead to new advances. Such was the case in the discoveries made by Galvani and Volta.

In 1700, Joseph, a Frenchman, and Caldani, an Italian, had noted the contractions of the muscles of a frog under the influence of electrical discharges. In 1752 Swammerdam, a Dutch physicist, published his account of an experiment in which a muscle was made to contract when touched at two points, one of which was a nerve, with the free ends of a silver and copper couple, the other ends of which were joined.

Johann Georg Sulzer, of Switzerland, published in Berlin, in 1762, the discovery which showed that when unlike metals such as silver and lead were held together on the tongue, a taste like that of iron sulfate was produced, and when one was held on the tongue and the other below it, there was no sensation until the outer edges of the metals came in contact.

GALVANI'S FROG EXPERIMENTS

Luigi Galvani (1737–1798), professor of anatomy at the University of Bologna, was not only a skilled anatomist but he was also familiar with chemical and physical sciences. During the year 1780, while he was dissecting a frog in his laboratory before a group of people including his wife Lucia (daughter of Domenico Galeazzi, Galvani's professor of anatomy) and Giovanni Aldini, his nephew, Galvani laid the frog on the table near an electrical machine that was being used at the time in performing electrical experiments. It was noted that when there was a spark, while at the same time a scalpel was held with its point at a nerve center, the legs of the frog contracted. According to one version of the story, it was Galvani's wife that called his attention to the unexpected happening.

Galvani became greatly interested in this rather astonishing phenomenon, especially since there was no direct connection between the machine and the scalpel or the frog. He did not attempt to explain the latter circumstance, but he recognized the fact that the contractions were due to electricity. Galvani tried various metals, other than the steel knife, with similar results. He also obtained the same results with a charged Leyden jar, with the electrophorus, and with atmospheric electricity. In the course of these experiments he had prepared a number of pairs of frog's legs which he mounted on brass or copper hooks, passing through the spinal cord, and had hung them upon an iron railing. He

Figure 3.1 Luigi Galvani's Frog-Leg Experiment *(Courtesy Burndy Library)*

found that there were contractions of the muscles when some part of the leg came in contact with the iron railing, not only during storms when there were flashes of lightning but also when there was a clear sky.

These observations led to further experiments in which Galvani used polished iron and copper hooks or rods, and also silver and tin. He was convinced that he was observing the effects of electrical charges somewhere in his apparatus and decided that the muscles or nerves themselves were the source of the electricity. He continued his experiments for eleven years before publishing, in 1791, an account of his observations. Before the publication of his experiments, however, others had learned of Galvani's discovery and repeated them.

VOLTA AND THE VOLTAIC PILE

Among those who became interested in this matter was Alessandro Volta (1745–1827), professor of natural history at the University of Pavia, whose name was mentioned previously in connection with the electrophorus. Volta was convinced by his experiments that the source of electricity was not in the nerves or muscles but in the metals. Galvani, however, showed that it was unnecessary to use metals at all. He found that contact of one of the nerves with the outer covering of one of the muscles was sufficient to cause contractions.

Galvani and Volta each drew to himself followers who supported the theories of their leaders and regarded as rank ignorance and heresy the theories of the other camp. The debate waxed hot and was never settled to the entire satisfaction of either side. Apparently there was truth on both sides, but it was Volta's ideas which finally led to important results.

Figure 3.2 Alessandro Volta *(Courtesy Burndy Library)*

Both Galvani and Volta continued their experiments, but Galvani died in 1798 and did not live to see the full fruition of his work. Aldini, who had worked closely with Galvani, and who had become professor of physics at Bologna in 1798, carried on Galvani's researches. He published several papers on the subject and became the chief protagonist of Galvani's cause. Volta may have invented the pile which bears his name as early as 1792, but the invention was not made public until March 20, 1800, at which time Volta addressed a letter on the subject to the Royal Society of London.

Volta's pile, or the voltaic pile, as it is more commonly called, took the form of a cylindrical stack, in which there was first a zinc disk, then a disk of felt, paper, or leather, soaked in a salt solution or dilute acid, then a copper disk, another zinc disk, another pad, and so on.

In his letter to the Royal Society, Volta described the behavior of the pile as similar to that of a feebly charged Leyden jar, but unlike that of the Leyden jar, the pile's charge was not dissipated but was constantly renewed. He observed that a spark was produced when wires connected to the two ends of the pile were brought together. Volta regarded this discovery as the experimental proof of his contention that the source of the electricity in Galvani's experiment was in the contact between dissimilar metals.

Volta also devised what was known as his *couronne des tasses* (crown of cups), in which strips of copper and zinc were hung over the edges of the cups and were partly immersed in a salt or acid solution. The copper strip in one cup was connected by means of a wire to the zinc strip of the next, and so on. When a connection was made between the zinc strip of the last cup and the copper strip of the first, a spark could be obtained, as with the pile.

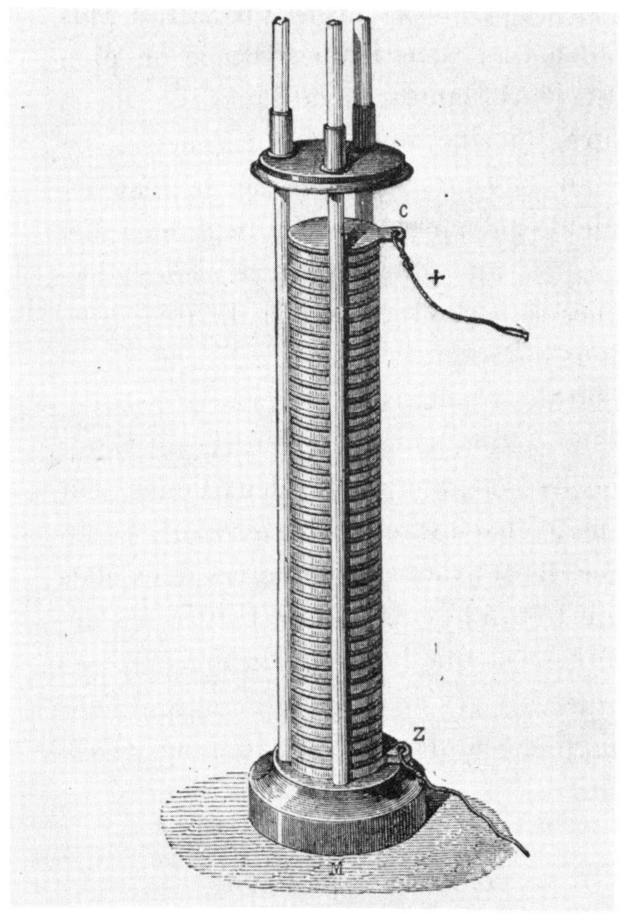

Figure 3.3 Volta's Pile *(Courtesy Burndy Library)*

Volta apparently believed in the contact theory, even after constructing his *couronne des tasses.* Other experimenters, however, went to work immediately, and by October 1800 Sir Humphry Davy had proved that the electricity was due to chemical action and that the voltaic cell would not operate with pure water. Carlisle and Nicholson discovered the decomposition of water on May 2, 1800, using a voltaic pile. In July 1800, William Cruikshank decomposed salts by similar means.

In November 1800, Volta went to Paris, where he lectured on galvanism and illustrated his lectures with experiments. Among other things he decomposed water with current from his cells. Napoleon recognized the importance of Volta's discovery and awarded him a gold medal, the cross of the Legion of Honor, and a prize of 6000 francs.

EVOLUTION OF THE BATTERY AND DISCOVERIES WITH ELECTRIC CURRENTS

In 1801 Johann Ritter developed the idea of a series of metals from which the relative electrical pressures produced by various pairs when immersed in a salt or acid solution could be determined. Volta conceived the same idea independently at a later date, and the series came to be known as Volta's electromotive series. These ideas indicated that the notion of electromotive force was already established, although a quarter of a century was yet to elapse before the enunciation of Ohm's law.

New developments came with great rapidity, and Volta himself had scarcely dreamed of the immense field that his discovery had opened. Cruikshank constructed a very powerful battery by soldering together plates of the same size of copper and zinc and using these bimetallic plates as separators between compartments in a trough. The trough was built of wood and was coated with pitch. In it were slots into which the separator plates were tightly wedged, and the intervening spaces were filled with a dilute sulfuric acid solution.

This battery produced very strong currents of electricity, with which it was possible to burn iron wires, heat charcoal to incandescence, and vaporize gold and silver leaf. The effects were, in many ways, much more powerful than those of the older electrical machines and Leyden jars, while in other ways the effects were not so great and less spectacular. The batteries produced only small sparks and had little effect on the nervous system; they did, however, produce very great heating and chemical effects. Since the electricities of the voltaic cells, and the older electric machines seemed totally unlike, the electricity obtained from the machines was for a time called common electricity while that of batteries was called voltaic electricity.

Experiments with the new voltaic cells were carried forward in many places. Sir Humphry Davy (1778-1829), one of the most brilliant scientists of his age, in his first lecture at the British Royal Institution, delivered on April 25, 1801, discussed the history of galvanism. In a paper presented in June of that year, he showed that galvanic action could be produced from plates of a single metal in different fluids, which were, however, in contact with each other.

Nicholas Gautherot, in 1801, discovered that when a current of electricity from a voltaic battery was sent through two copper plates in sulfuric acid, for a short time these plates became capable of supplying a current in the opposite direction. Ritter, between the years 1803 and 1805, constructed several secondary batteries using copper, gold, and other metals as electrodes. In 1802 Luigi Valentino

Brugnatelli, a friend and pupil of Volta, discovered electroplating, with the use of the voltaic pile.

Volta's first pile was a puny affair, and the immediate reaction of other scientists, who became interested in the matter, was to build larger and still larger batteries. Several learned societies, among them the Royal Society, appropriated funds for the construction of such batteries, some of which were huge things, weighing tons.

William Hasledine Pepys constructed several very large batteries for the Royal Society between the years 1802 and 1808, using copper and zinc plates in a nitric acid solution. Dr. Wollaston also built several huge batteries, and Dr. Robert Hare of the University of Pennsylvania constructed

Figure 3.4 Wollaston's Pile. This great electrochemical pile was built by the Royal Institution in London with funds provided by public subscription in 1813. It was intended as a rival to the pile of 600 couples built by the order of Napoleon in Paris and was used by Humphry Davy for his experiments. *(Courtesy Burndy Library)*

batteries strong enough to fuse large pieces of metal. There were also batteries of great size at Paris and smaller ones at most of the universities of Europe.

Nearly all of these batteries were constructed of the same simple elements, and they were all subject to the defect called polarization. Polarization refers to the collection of hydrogen gas bubbles on the positive electrode, causing the internal resistance of the cell to increase, and setting up a counter electromotive force, thereby diminishing the output.

After Davy had proved the chemical nature of the voltaic cell and other experimenters had succeeded in decomposing water, there developed a great interest in the chemical action of the voltaic current. Berzelius of Gotland, Sweden, between the years 1802 and 1806, published a number of important papers on his electrochemical researches. These papers undoubtedly laid the foundation for the brilliant series of electrochemical discoveries by Davy.

On November 20, 1806, Davy delivered his first Bakerian lecture before the Royal Society, on the subject, "On Some Chemical Agencies of Electricity," and thereby provided the basis for the ionization theory. Then followed that notable series of discoveries in which Davy isolated sodium and potassium in 1807 and barium, boron, calcium, and strontium in 1808.

Funds for a powerful new battery for the Royal Institution were subscribed in 1809. With this battery Davy separated the halogens, iodine, chlorine, and fluorine. These elements had, however, already been isolated by others. In 1810, Davy for the first time exhibited the carbon electric arc, using the battery as a source of electricity.

Figure 3.5 Sir Humphry Davy *(From Smithsonian Institution)*

ELECTROMAGNETISM

Hans Christian Oersted (1777–1851) was professor of natural philosophy at the University of Copenhagen. As early as 1812 he expressed the belief that magnetic fields were associated with electricity, but this statement was by no means a new doctrine, for electricity and magnetism had long been considered as having something in common, probably as far back as Thales. There was already much evidence that the two were related, such as the fact that pieces of steel had been magnetized by strokes of lightning and steel needles had been magnetized by discharges from electrical machines. Boze, in a letter to the Royal Society, stated that he had reversed polarities in a magnet and had destroyed magnetism by means of electricity. Beccaria had commented that the earth's magnetism might be due to the circulation of the electric fluid about the earth. Franklin knew that lightning produced magnetic effects.

Romagnosi, in 1802, had observed the deflection of a magnetic needle in the presence of a conductor carrying an electric current, but he seemed to have attached no importance to his discovery at the time, and certainly no results came of it. Romagnosi himself made no claims to priority in the discovery of electromagnetism.

In 1819, Oersted was lecturing before a science class where he demonstrated the heating effects of voltaic electricity. During the demonstration he noted the oscillations of a nearby compass needle, which corresponded with the opening and closing of the electrical circuit. Oersted spent some months with further experiments on this phenomenon, using stronger batteries in order to obtain greater deflections of the needle. He found that not only was a compass needle deflected by the electric current, but a wire

Figure 3.6 Oersted's Discovery *(Courtesy Burndy Library)*

carrying a current of electricity, when free to move, was deflected by a magnet.

Oersted's great discovery was published in a Latin monograph dated July 21, 1820. The effect on the scientific world was like that of von Guericke's electric machine or von Kleist's Leyden jar, for it was news of the greatest importance, and Oersted's experiments were repeated by men of science everywhere.

AMPÈRE

André Marie Ampère (1775–1836) discovered that there were forces between two wires carrying electric currents. He supported a loop of wire with its terminals dipping into mercury cups. When a current of electricity was sent through this loop and an adjacent stationary loop, the suspended loop was deflected through an angle. Between September 18 and November 2, 1820, Ampère delivered a series of lectures at the French Academy, describing new discoveries in electromagnetism. He defined the relationship between the direction of current flow and the deflection of a magnetic needle, and he showed that parallel conductors carrying currents in the same direction attracted each other, while those carrying currents in opposite directions were mutually repelled. He also constructed a lightly suspended helix of wire and showed that when an electric current passed through it, the helix behaved like a compass needle and like a magnet with respect to other magnets. He developed the astatic needle, in which the effect of the earth's magnetism is neutralized, advanced the theory that the magnetism of permanent magnets might be due to electric currents in their ultimate particles, and suggested that electromagnetism might be used for telegraphy.

ARAGO, BIOT AND SAVART

Arago discovered, in 1820, that a wire carrying a current of electricity would attract iron filings. He and Ampère both discovered that a wire helix carrying an electric current was capable of magnetizing a soft iron bar around which it was wound. In the same year Biot and Savart announced the law governing the force between a long straight conductor and a magnetic pole which is expressed by the equation $H = 2i/r$, in which H is the magnetic field strength, i is the current, and r is the perpendicular distance from the point of measurement to the conductor that carries the current.

FARADAY'S ROTATING CONDUCTOR AND MAGNET AND BARLOW'S WHEEL

On September 3, 1821, Michael Faraday discovered that a conductor carrying a current would rotate about a magnetic pole and that a magnetized needle would rotate about a wire carrying an electric current. For the first experiment, he used a metal cup containing mercury in the center of which was a vertical bar magnet, one of whose poles protruded above the surface of the mercury. The conductor was lightly suspended above the cup, with the lower end dipping into the mercury. When a current was sent through the conductor, it revolved about the magnetic pole. For the second experiment, Faraday used substantially the same equipment except that the conductor was held in a fixed position, touching the mercury at the center of the cup, and the magnet was a magnetic needle, one end of which was weighted with platinum, so that one pole rested upon the bottom while the other end floated freely in the mercury, with the pole projecting above the surface of the mercury. When a current was sent through the conductor into the mercury, the magnetic needle rotated about the conductor.

At about the same time Faraday discovered that a metal disk mounted on an axis through its center with its lower edge placed between the poles of a horseshoe magnet would revolve when a current passed between the axis and a trough of mercury into which the lower edge dipped. Peter Barlow had made a similar motor a little earlier, and the device was therefore called Barlow's wheel.

STURGEON'S ELECTROMAGNET

Sturgeon is credited with making the first electromagnet in 1821, although Ampère and Arago had noted the magnetic fields of helices and their ability to magnetize soft iron. Professor Joseph Henry, of Albany, New York, was the first to make electromagnets with superimposed layers of insulated wire. His magnets were of high intensity and great lifting power.

GALVANOMETERS

Schweigger and Poggendorff, during the years 1820 and 1821, constructed the first crude galvanometers, using a wire helix and a compass needle. The earliest instrument of that type was called a multiplier, because the effect of a single wire on a compass needle was multiplied many times by the use of a helix.

AMPÈRE'S AND OHM'S LAWS

Except for Coulomb's torsion balance, previous measuring instruments were scarcely more than indicators. Improvements were soon made in the Schweigger and Poggendorff galvanometers so that they became measuring instruments of great accuracy. After Ampère published his fundamental law governing the relationship between an electrical circuit and a magnetic field, it became possible to use the galvanometer for accurate measurements of electric currents. This law, which was published in 1822, is as follows: $f = i \cdot m \cdot ds/r^2$, in which f is the force produced by a current i and a magnetic pole strength m; ds is an incremental length of conductor, which is perpendicular to the line joining the conductor and the magnetic pole and separated from it by a distance of r centimeters. From this law may be derived a definition of the absolute unit of current strength.

Georg Simon Ohm (1787-1854) began his study of electric currents, flowing through conductors, in 1825. He published the results of his studies in 1827, and in his paper, the law, which now bears his name, was presented for the first time. In present-day international symbols, this law reads $I = V/R$, where I is the current, V is the electromotive force or voltage, and R is the resistance.

The heavy curtain, which for so long had obscured in mystery the nature of electricity and magnetism, had been pulled aside just a little, but most of nature's great secret was still hidden. There were at hand, however, two new prophets in the temple of science, who soon were to open the curtain much wider.

4

Faraday and Henry

FARADAY'S FORMATIVE YEARS

Michael Faraday was born at Newtington Butts, South London, September 22, 1791. His father, James Faraday, was a journeyman blacksmith. Michael's formal education ended at the age of thirteen, at which time he took employment on a trial basis with a Mr. Riebau, a bookseller, in Blandford Street. After a year he became apprenticed to Mr. Riebau for a period of seven years. During his apprenticeship, Faraday became a skilled bookbinder, and at the same time he developed an absorbing interest in the sciences, and particularly in chemistry. His bent toward science had its origin in certain books, which passed through his hands in the Riebau shop.

As a youth Faraday read widely, not only in the sciences but also in other fields of literature. He developed a certain facility and beauty of expression that in later years became one of his great assets. Faraday's career in experimental science began with the acquisition of a few pieces of chemical apparatus. He attended several courses of popular scientific lectures and there made the acquaintance of other young men whose interests were much like his own.

The event that changed the whole course of Faraday's life was an invitation from a Mr. Dance, who visited the Riebau establishment frequently, to attend four lectures by the then famous Sir Humphry Davy. These lectures were delivered on February 29, March 14, April 8, and April 10, 1812. Faraday made careful notes of the subject matters, which he later transcribed, and then bound into a book.

It was at about this time, when Faraday was twenty-one years of age, that he completed his apprenticeship with Mr.

Figure 4.1 Michael Faraday *(From Smithsonian Institution)*

Riebau. Notwithstanding a deep affection for his master, Faraday had developed a dislike for his trade, but nevertheless he took a position as a journeyman bookbinder with a certain De la Roche. The disagreeable nature of his new employer caused Faraday to reach the definite decision that he had no desire to remain a bookbinder. He confided his longing for scientific work to Mr. Dance, who urged him to write to Sir Humphry Davy. Faraday followed his friend's advice, wrote a letter, and sent along the bound notes of Davy's lectures. Davy was impressed, and asked Faraday to come to him for an interview.

FARADAY APPOINTED TO THE ROYAL INSTITUTION

Davy urged upon Faraday the thought that science was a hard and penurious master, but Faraday was not dissuaded. Davy apparently liked the young bookbinder, for he recommended his appointment as a laboratory assistant. Faraday was given the position at a salary of twenty-five shillings a week, together with living quarters at the Royal Institution.

The appointment was made in March 1813, and on October 13, 1813, Faraday left London to accompany Sir Humphry and Lady Davy on an extended tour of the Continent. They visited France, Belgium, Italy, Switzerland, and Germany, and in the course of their travels they met Ampère, Arago, Gay-Lussac, Chevreul, Dumas, Volta, de La Rive, Biot, de Saussure, de Staël, Count Rumford, and Humboldt. Faraday gained greatly in knowledge and was very happy, except for the unpleasantness caused by Lady Davy who treated him as a servant.

They returned to England on April 23, 1815, and two weeks later Faraday was again engaged at the Royal Institution, as a laboratory assistant at a salary of thirty shillings per week. During several years that followed, he worked at the side of Sir Humphry in the laboratory, and also assisted him at the lecture table. Faraday aided Davy in the development of the miner's safety lamp and in many of the experiments that Davy was pursuing, but at the same time Faraday carried on a certain amount of original research. In 1816 he presented a series of seven lectures before the City Philosophical Society on chemical subjects. Although he was becoming known in a limited way, he was so close to the brilliant Davy that for a time his talents were obscured.

On June 12, 1821, Faraday married Sarah Bernard, who was a member of a religious sect called Sandemanians, of which Faraday was also a member. Their marriage appears to have been very happy for both, throughout their long lives.

When Oersted's discovery of electromagnetism was announced in 1820, Davy brought a copy of the paper to the laboratory immediately, where he and Faraday repeated the experiments. Faraday was occupied at the time with many special problems, including alloy steels, but he seems to have been much impressed with Oersted's discovery and during his leisure moments returned to the experiment many times. He made his first independent discovery after his marriage, namely, the rotation about a magnetic pole of an electrical conductor carrying a current. This discovery came on September 3, 1821, and on Christmas Day of the same year he showed his wife a conductor carrying a current rotating under the influence of the earth's magnetism. During the year 1821, which had been an eventful one for Faraday, he was made superintendent of the Royal Institution.

In addition to his lectures, experiments, administrative duties, and researches for the government, Faraday began to receive many requests for special research work, of a private nature, some of which he undertook. The fees he received for this work might have made him wealthy, but some years later he gave up such work altogether, because he felt that it was interfering with what he considered more important projects.

Faraday's mind reverted constantly to electrical and magnetic problems. In 1822 he had written in his notebook the words, "convert magnetism into electricity." During the years that followed, Faraday returned to this problem again and again. If his apparatus had been better, the discovery would probably have been made in 1824 or 1825. In the meantime important work in other fields was bringing him recognition. He had liquified a number of gases, including chlorine, and had read a paper on the subject in 1823. In 1824 he produced benzol, or benzine, and during that year he also began his researches on optical glass, which continued until 1829.

In 1825 Faraday was appointed to the post of director of the laboratory and professor of chemistry at the Institution. In 1826 he began his Friday evening lectures, which were to continue for many years, and during that year also, he began his Christmas lectures for children. On December 3, 1827, Faraday took as his assistant Sergeant Anderson who, until his death in 1866, worked faithfully at Faraday's side.

ELECTROMAGNETIC INDUCTION

In 1824, Arago had noted that a certain very fine compass, which had had copper in the bottom of its cell, oscillated for only a short time when in its cell but when the needle was removed it oscillated over a longer period. The copper was tested and was found to contain no iron. Arago then tried rotating a copper disk under a magnetic needle and found that the needle tended to follow the rotation of the copper disk. It was found also that other metals exhibited similar effects. Silver showed a greater effect than copper, while lead, mercury, and bismuth were less effective. A copper disk was made to rotate by revolving a magnet near it. Faraday and other researchers were aware of these phenomena but none could offer a satisfactory explanation.

In September 1831 Faraday again attacked the problem of producing electricity from magnetism. His experiments at this time are described in his *Experimental Researches in Electricity*. His first apparatus consisted of a wooden spool, on which were wound twelve separate helices, insulated from one another with calico and the individual turns separated by a winding of twine. He connected the even numbered coils into one series and the odd numbered coils into another. The one set of coils he connected to a battery and and the other set to a galvanometer. With a current flowing from the battery, he saw no effect on the galvanometer. Later, with a much more powerful battery, consisting of a hundred cells, connected to a single pair of insulated coils, he noted slight effects on the galvanometer when the circuit was made or broken. The deflection was in one direction when the circuit was made and in the other direction when it was broken. As long as there was a steady flow of current from the battery, the galvanometer showed no effect.

His next experiment is generally considered as the one in which he proved conclusively that electricity could be produced from magnetism. He used a welded iron ring 6 inches in outside diameter, of 7/8 inch soft iron rod. On one sector of the ring he wound three coils, covering about 9 inches of the circumference, and on another part he wound a single coil containing about 60 feet of copper wire, about 1/20 inch in diameter. It should be noted that commercially insulated magnet wire did not exist until about twelve or fifteen years later.

In Faraday's description of his experiment, he designated the triple coil as "A," and the single coil as "B." The coil "B" was connected to a galvanometer, and the three coils were joined together, with the terminals connected to a battery through a switch. When the battery circuit was closed there was a powerful deflection of the galvanometer in one direction, and when the circuit was opened there was an equally strong deflection in the opposite direction.

In his next experiment, Faraday dispensed with the battery and instead made use of a very large compound magnet. He mounted a helix on an iron bar, which extended out beyond the helix on either side. As the bar carrying the coil was brought near the magnet, it was drawn suddenly to the poles. The galvanometer, which was connected to the coil, was deflected violently as the bar was attracted to the magnet, so that the needle spun around. When the bar was pulled suddenly away from the magnet, an equally violent deflection of the galvanometer needle occurred in the opposite direction.

A single turn of copper strip about the iron bar was sufficient to produce oscillations of the galvanometer needle. When the helix without the iron bar was moved across the field of the magnet, there were wide swings of the galvanometer needle, but the effect was much weaker than with the iron core.

Faraday made many variations of these experiments, in one of which he produced a steady current through a copper disk when it was rotated between the poles of a magnet. The galvanometer connection was made between the axis of the disk and its periphery, with a sliding contact on the latter. This device will be recognized as a unipolar generator and the reverse of Barlow's wheel described in the previous chapter.

Faraday used the current obtained by induction from his coils to produce a spark between charcoal points, and with the discharge from the same coils he magnetized needles. He also constructed a simple loop of wire, whose terminals were connected to two commutator segments, upon which two brushes rested. When the loop was turned upon its axis in a magnetic field, his galvanometer indicated a pulsating, unidirectional current.

It now became clear to Faraday what the significance was of the damped swings of Arago's compass needle and the revolution of the needle when placed near a revolving copper disk. Induced currents in the copper disk produced magnetic fields that opposed the relative motions of disk and needle.

Most of this series of experiments had required the short space of only a few days, and in a few weeks Faraday had a pretty complete notion of the processes of electromagnetic induction. His first paper on the subject was delivered to the Royal Society on November 24, 1831, entitled "Experimental Researches in Electricity." News of the discovery caused a tremendous reaction throughout the world. The way now lay open for the conversion of mechanical energy into electrical energy, but the production of practical machines, by means of which this much-desired end was to be accomplished, still lay in the future.

OTHER CONTRIBUTIONS BY FARADAY

The discovery of electromagnetic induction was one of the greatest discoveries of all time in its effect on the material welfare and progress of mankind. Faraday's work in electricity did not, however, end here, for he made great strides in electrochemistry, determined the specific inductive capacities of many materials, established a relationship between light and magnetism, investigated diamagnetism, and with the assistance of his friend, the Reverend W. Whewell, established an important part of our present-day electrical nomenclature. Faraday brought into common use the following terms:

electrode	ion	diamagnetic
anode	anion	dielectric
cathode	cation	electrochemical equivalent
electrolyte	lines of force	specific inductive capacity
electrolysis	paramagnetic	

Berzelius must be considered the father of electrochemistry, and although Davy made many great new discoveries in this field, it was Faraday who established it as an exact, quantitative science. He developed the theory of ionization and discovered the relationships from which he determined the electrochemical equivalents of a large number of elements.

Faraday gave up his private practice about the year 1832. He had accumulated some money, and his needs were relatively simple. He did, however, continue to serve the British government in many ways. He became a lecturer at the Woolwich Royal Military Academy in 1829 and continued in this capacity until 1849. For a number of years he was a consultant to Trinity House on lighthouses, and in 1847 he proposed lighting buoys with incandescent lamps using platinum spiral filaments.

No other man contributed so much to electrical science as did Faraday, and his achievements were recognized during his lifetime by many learned societies and universities in many countries. He was humble, kindly, deeply religious, quiet, persevering, and had almost an intuitive insight into many of nature's secrets, and beheld the wonders of creation with a reverent awe. His life ended in 1867.

JOSEPH HENRY

There is an amazing similarity between Faraday's life and works and those of his American contemporary, Joseph Henry. The two men were very much alike in many respects. Henry was born at Albany, New York, either on December 17, 1797, or 1799, probably the latter. The confusion is due to the fact that the record in the old register at Albany is not very distinct. He was of Scottish ancestry. At an early age he went to live with his maternal grandmother at Galway in Saratoga County where he attended the district school. At the age of ten he began working half days at a store in the village and attended school only part-time. While he was still at Galway, he gained access to a library in the village church, and as a result he became a passionate reader.

At the age of fifteen, Joseph left Galway and returned to his mother's house in Albany. His father had died some years earlier, and his mother supported herself and her son by keeping boarders. For a time, after his return to Albany, it seemed that Joseph might become an actor. He had considerable talents in acting, in arranging scenery, and in playwriting, but his interests were suddenly shifted in another direction. At the age of sixteen, he was confined to the house for a brief period due to an accident. While he was so confined, he happened upon a book belonging to a gentleman who was then a boarder at his mother's house. The book was entitled *Lectures on Experimental Philosophy*,

Astronomy, and Chemistry by Dr. Gregory. This book so fascinated young Henry that he resolved, forthwith, to devote his life to serious study.

Upon his recovery, Joseph Henry appeared before the other members of "The Rostrum," a dramatic organization of which he was then president, and tendered his resignation. His next step was to enter a night school, whose facilities he soon exhausted. He then took up English grammar and rhetoric with a private teacher. These studies were followed by a brief period of tutoring, after which he entered Albany Academy as an advanced pupil. In order to meet his expenses, he began teaching in a neighboring school. Next he became an assistant at the Academy, and a little later he accepted a position in the family of S. Van Rensselaer as a private tutor. While he was in the Van Rensselaer household, he had much leisure time, which he devoted to studies in the subjects which interested him most.

He was at this time considering preparing himself for the medical profession, but it happened that he was offered a position as a surveyor with a party engaged in laying out a new road in southern New York, and he accepted the offer. A few months later, after the survey had been completed, Henry was offered a professorship at Albany Academy, which offer he quickly accepted.

HENRY'S FIRST EXCURSIONS INTO SCIENCE

The great series of electrical experiments and discoveries that soon brought fame to Joseph Henry began here at Albany Academy. For an American city of that period Albany was well suited to the development of scientific pursuits, because there were several societies whose purpose was to promote scientific experiments and discussion.

Dr. Beck, who was president of Albany Institute, which

was one of these organizations, presented a series of lectures on chemistry, at which he was assisted by Henry. Unfortunately Henry's duties as a professor occupied so much of his time that he had little leisure for experiment. His teaching required seven hours of each day, eleven months of the year.

Henry's earliest scientific papers dealt with the adiabatic expansion of steam and with the cooling effects produced by the expansion of gases. At this time, he was also assisting in compiling meteorological data for the Albany area.

His first electrical paper, delivered before the Albany Academy on October 10, 1827, dealt with the subject of electromagnetism, and in this paper he described an electromagnet made by William Sturgeon of London in 1825. Henry described this magnet as horseshoe shaped, wound with a single widely spaced layer of eighteen turns of wire over a varnish-insulated iron core. With a powerful battery this magnet was capable of lifting 9 pounds. Sturgeon's magnet was the first to employ an iron core. Ampère and Arago in France had made helices that behaved like magnets in 1820 and 1821. Johann Schweigger of Halle, Germany, aided by J.C. Poggendorff, built their so-called multiplier in 1820, in which closely spaced coils of wire insulated with fabric were wound upon a nonmagnetic spool.

By combining the ideas associated with the Sturgeon electromagnet and the Schweigger multiplier, Henry began the construction of electromagnets in which he was the first to use superimposed coils of insulated wire, wound upon horseshoe-shaped iron cores. The insulation, which was applied by hand, was cotton or silk. For a later electromagnet it was said that Henry used a silk dress belonging to his wife. In 1829, he demonstrated an electromagnet using 400 turns of wire with a lifting power of many pounds, but the exact figure is not known nor is the power of his battery. Later in the same year he completed another magnet, with a ½-inch iron core wound with several coils. He believed it was necessary to distribute the magnetizing force along the core. Because his electromagnets had coils with many turns, he called them intensity magnets. This latter magnet, with a small battery, supported a weight of 39 pounds.

HENRY PROPOSES THE ELECTROMAGNETIC TELEGRAPH

After constructing a number of electromagnets, Henry became interested in operating an electromagnet at a distance. In 1830 he connected one to the end of a line 1030 feet long and was able to operate it successfully from the other end. Henry pointed out that it was possible to construct a telegraph in this way.

With the aid of P. Ten Eyck, he next built an electromagnet with a core weighing 21 pounds, wound with nine coils, each of which contained 60 feet of wire, insulated with waxed cotton. This magnet supported 750 pounds. In 1831 at Yale College Henry supervised the making of an electromagnet that supported 2065 pounds.

ELECTROMAGNETIC INDUCTION

Unlike Faraday, Henry did not keep a notebook to record his ideas and experiments, nor did he write many papers at this period of his life. In a later paper he mentions an experiment, which he conducted in the early 1830s, in which he discovered independently the principle of electromagnetic induction, but Henry credited Faraday with the discovery. The apparatus used by Henry consisted of an iron bar in the middle of which was a coil connected to a galvanometer. He placed the iron bar carrying the coil across the poles of an electromagnet connected to a battery and found that when he raised and lowered the plates of the battery into and out of the electrolyte, the galvanometer needle was deflected, first in one direction and then in the other.

SELF-INDUCTION

While Faraday deserves the credit for the discovery of electromagnetic induction, it was Henry who first discovered self-induction. Self-induction is that property of an electrical circuit that causes a counter electromotive force when the circuit is made or broken and is designated by the symbol *L*. A little later, Faraday discovered self-induction independently. Henry was honored in the use of his name for the unit of self-induction.

MARRIAGE AND PROFESSORSHIP AT PRINCETON

Henry had now become famous, but he remained a poor professor, and while money matters were of little concern to him, the lack of money undoubtedly hampered him in his experimental pursuits. Despite his poverty and the improbability that he could ever hope to earn a comfortable living, he had married Harriet L. Alexander, his cousin, in 1830. The marriage appears to have been a happy one in which six children were born to them, three of whom survived Henry.

Near the end of the year 1832, Henry was offered the chair of Natural Philosophy at Princeton University, which he promptly accepted. As was the case at Albany, his teaching duties were arduous and afforded him little time for experiment. In addition to his regular courses, which in modern terms would probably be called physics, he also taught chemistry, minerology, geology, astronomy, and architecture. His burdens of pedagogy notwithstanding, he found time to construct another magnet with a lifting power of 1 ½ tons. He also installed a telegraph signal, including a relay, between the college and his house, a distance of about a mile, using the earth as one side of the circuit.

During what were rather happy years for him in the academic surroundings of Princeton University, he continued his experiments in electromagnetism, electromagnetic induction, and self-induction. He was loved and revered by his students and was regarded as perhaps the foremost scientist in his field in America.

In 1837, the university sought to reward Henry in some measure for the extraordinary services he had rendered, both to the university and to the world at large. He was granted a one-year leave of absence to visit Europe, on which visit he was accompanied by A. D. Bache, a descendant of Benjamin Franklin. Henry was received with great courtesy by European scientists, including Faraday, Wheatstone, Daniell, Arago, and de La Rive. He was also invited to speak before gatherings of the learned societies.

ELECTRICAL OSCILLATIONS AND ELECTROMAGNETIC WAVES

Refreshed by his sojourn in Europe, Henry returned to his post at Princeton and to his experiments. He had discovered the essential principles of the transformer and experimented with coils placed at varying distances from each other. He also discovered the oscillatory nature of the discharge of a Leyden jar by discharging one across a gap through a circuit containing a small coil. In this coil he placed an unmagnetized steel needle and found that after discharging the Leyden jar the needle was magnetized, sometimes in one direction, and again in the opposite direction.

This small coil and needle became very important in other experiments, as for example in detecting the inductive effects of lightning, which he accomplished by connecting the coil between the metal roof of his house and the ground. Lightning flashes up to 20 miles away caused the needle to be magnetized.

In a somewhat similar experiment, he made a discovery whose great importance was not recognized until fifty years later. For this experiment, Henry strung a circuit of wire about the perimeter of the upper floor of a building, and a similar closed circuit, containing the coil and needle, around the walls of the cellar. The two circuits were about 30 feet apart. When a Leyden jar was discharged through the circuit on the upper floor, he found that the needle in the coil in the cellar had been magnetized. From his description of this experiment it is clear that Henry recognized that he was dealing with a phenomenon different from the electromagnetic induction experiments, in which the coils were close together and connected by magnetic circuits containing iron. He compared the effects in the present experiment with those of light and spoke of the disturbance being propagated wave-fashion.

OTHER RESEARCHES

While still a professor at Princeton, Henry carried on researches in many other fields, notably in meteorology, but also in metallurgy, ballistics, light, and astronomy. The congenial atmosphere at Princeton held Henry for fifteen years.

THE SMITHSONIAN INSTITUTION

In the year 1837 the United States government received a bequest of about $500,000 from the estate of James Smithson, an Englishman, for the purpose of founding "at Washington, under the name of the Smithsonian Institution, an establishment for the increase and diffusion of knowledge among men." For ten years the bequest was tossed about in Congress, but an organization was formed and a board of regents appointed. After so many years of deliberation the board of regents, in 1847, asked Joseph Henry to become the director of the Smithsonian Institution and to undertake its organization.

The founding of the Smithsonian Institution brings up a very interesting parallel. The Royal Institution in London was founded in 1799, and the moving spirit in its founding was Benjamin Thompson (Count Rumford), an American by birth, who had a most colorful career in Austria, England, and France. It was he who first appointed Sir Humphry Davy as the head of the Royal Institution. The Smithsonian Institution was founded by an Englishman, but it was Joseph Henry who gave it purpose and direction.

The offer of the regents was accepted by Henry, whereupon he resigned his professorship at Princeton and moved to Washington. He presented an outline of his conception of the organization and purposes of the Institution which the regents accepted. For thirty-two years Henry guided the Smithsonian Institution, writing, administering, planning, and experimenting. Under Henry's administration, the original bequest was not, as might have been expected, dissipated but rather increased to more than double its original value.

Despite the great burden of administrative duties connected with the directorship, Henry found time to carry on

Figure 4.2 Joseph Henry *(From Smithsonian Institution)*

experiments of the most varied kinds, including acoustics, strength of materials, testing of munitions, advising on military inventions, operation of lighthouses, foghorns, collecting meteorological and astronomical data, and many other things. Henry's well-filled life ended at Washington, D.C., on May 13, 1878.

5

Direct-Current Dynamos and Motors

After the discoveries of Oersted, Faraday, and Henry in electromagnetism, electromagnetic induction, and self-induction, all of the necessary facts needed for the development of the generation of electricity by mechanical means were known, but the construction of actual, practical machines still required the solution of many problems. The purely research aspects of electricity were now blending into engineering problems.

PIXII'S MACHINE

The first small machine was built in 1832 by Hypolite Pixii of Paris, an instrument maker who had become interested in electricity. This machine was, in modern terms, a magneto and consisted of two coils on a U-shaped core, the pole faces of which were opposite the poles of a revolving permanent magnet. In its original form the machine produced alternating current, but Pixii also built machines in which the coils revolved and which were equipped with commutators. In the same year, Dal Negro, an Italian, built a similar magneto electric machine independently of Pixii.

Many others became interested in building small magneto machines over the next eighteen years, including Saxton of Philadelphia in 1833, E. M. Clarke of England in 1834, and Page of Washington in 1835. During this period, there were probably vast numbers of other such machines built by unknown scientists, professors, mechanics, and others.

NOLLET'S MACHINES

The first improvement, after Pixii's machine, was made by M. Nollet, a professor of physics in the military school at

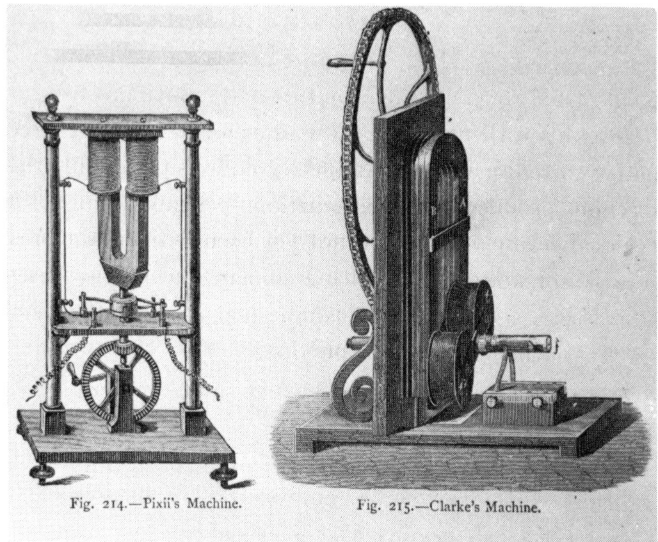

Figure 5.1 Pixii's Machine and Clarke's Machine. Pixii's machine for generating an electric current had a rotating magnet that induced alternating magnetic fields in the soft iron cores and an electric current in the coils that surrounded them. The current reversed twice per revolution but was converted to direct current by a commutator. Clarke's machine employed the same principle but had rotating coils and a fixed magnet. *(Courtesy Burndy Library)*

Brussels. With the aid of his assistant, Joseph Van Malderen, he built a large machine in which thirty-two permanent magnets were mounted on a stationary frame in longitudinal rows, spaced at equal angular distances. The rotor within this frame consisted of a number of coils equal to the number of pole faces of the permanent magnets on the stator.

The magnet poles were, of course, arranged so that they were alternately north and south, and the coils were connected in such a manner that the induced electromotive forces became additive. The terminal wires were connected to insulated rings mounted on the shaft, which had brushes bearing upon them in much the same manner as later alternating-current machines. At 400 revolutions per minute this machine produced 6400 alternations per minute, or 53¹/₃ cycles. Because voltage had not yet been defined, the pressure is not known, but since a similar machine was used later for arc lights, we may assume that it was in the vicinity of 50 volts or perhaps a little more.

The exact date of Nollet's machine is indefinite, but it was probably about 1849 or 1850. In 1853 a company was organized in Paris for the purpose of manufacturing machines on the Nollet design. This company, called the Alliance Company, was probably the first electrical manufacturing concern in the world. It produced several sizes of machines, having from thirty-two to forty-eight magnets, and from sixty-four to ninety-six coils. Originally the Alliance machines were to be used for electrochemical work and, what now seems stranger still, for the decomposition of water to obtain oxygen and hydrogen for use with the limelight or calcium light. For this purpose the machines were equipped with commutators. Unfortunately, the use of the machine for the decomposition of water was a business failure.

F. H. Holmes, of England, adapted the Alliance machine to arc lighting, using carbon points. Motive power was furnished by a steam engine, and it was found that it required 5 horsepower to supply the electrical energy for one 1100-candlepower arc lamp. This and similar machines were soon put to work in several British lighthouses and Faraday himself directed the first installation at South Foreland in 1858.

DYNAMOS

In 1845 Wheatstone suggested the use of electromagnets in place of permanent magnets for the fields of the new machines, and similar suggestions came from Sören Hjorth of Denmark and Varley of England. The first such machine was built by H. Wilde of Manchester. It was not self-exciting, but instead it was equipped with a small magneto used as an exciter. This arrangement was eminently successful, so much so that the machine overheated because of overload and also because of eddy-current losses in the solid iron cores of the armature and field magnets.

Wilde's machine employed the shuttle armature, which had been invented a short time previously by Dr. Werner Siemens of Berlin. The success of this machine was followed shortly by the construction of self-exciting machines at almost the same time by Moses G. Farmer of Salem, Massachusetts, Alfred Varley and Charles Wheatstone of England, and Dr. Werner Siemens of Berlin. These were called dynamoelectric machines, or simply dynamos, as contrasted with the magnetoelectric machines, or magnetos.

Antonio Pacinotti of Florence, Italy, developed a slotted ring armature in 1860, which was the prototype of most later machines. Zénobe Théophile Gramme of Belgium, who had been a patternmaker for the Alliance Company, in 1871 constructed a ring armature, in which the coils were wound on a smooth iron ring. Although the Gramme armature was a more recent development, it was later abandoned in favor of the Pacinotti type.

ELECTRIC MOTORS

Many experimenters had sought to convert electrical into mechanical energy or, in other words, had sought to build an electric motor. Several such devices have been mentioned previously in this work, such as the whirligigs of Wilson and Hamilton, Franklin's electric wheel, Barlow's wheel, and Faraday's revolving conductor, but none of these was of any practical value.

In 1832 Sturgeon built a small electric motor of a general type, of which there were many variations, in which contact is made at regular intervals to energize an electromagnet, causing it to attract a revolving iron bar as it approaches the magnet's poles. Henry built a walking-beam motor on this principle. The most noteworthy attempt of this kind was made by Thomas Davenport, a blacksmith of Brandon, Vermont, in 1834. He built such a motor powerful enough to operate a small electric railway, and in 1840 he operated a printing press with the power supplied by one of his motors.

Professor Jacobi of St. Petersburg built a somewhat similar motor in 1834, and in 1839, with the use of such a motor, propelled a boat carrying fourteen passengers. A Scotchman, Robert Davidson, in 1838 built an electric car weighing 6 tons. There were without doubt hundreds of other similar attempts, but until the last quarter of the nineteenth century there were two serious drawbacks: there was as yet no satisfactory motor, and there was no cheap source of electric power.

In 1873, at an industrial exposition in Vienna, there occurred one of those fortunate accidents that profoundly affected the course of electrical development. At the exposition was an exhibit of Gramme dynamoelectric machines. While one of the machines was in operation, a workman carelessly connected the terminals of another machine to the wires that carried the electrical energy from the first machine. The second machine began to whir and spark at the commutator and attained a considerable speed before it could be disconnected. When Gramme was informed of this startling event, he grasped its importance immediately. The news was carried to the world through the technical journals of the day and created a great stir.

Within the next decade the electric motor had been put into service in many applications for which a compact source of power requiring little attention was desirable such as railways, elevators, mine hoists, machine tools, printing presses, fans, sewing machines, and grinders.

Improvements in Batteries and Electrostatic Machines

The development of the electric dynamo and motor occurred during a period when there were many improvements in older kinds of electrical devices. The field of electrical research had broadened greatly, and no single line of inquiry was any longer sufficiently important to merit the entire attention of electrical experimenters.

During the period under consideration many new forms of voltaic cells appeared. In the first thirty years after Volta, practically all batteries were simple copper and zinc couples in an acid solution. Volta himself had compiled a list of elements, known as the electromotive series, which indicated the electrical pressures that could be developed between various couples. Since Volta had no measuring instruments worthy of the name, such an arrangement was an achievement of genius.

The original copper, zinc, and acid batteries had several serious faults, chief among which was polarization, which has been mentioned previously. Because of this defect, it was impossible to draw a large current from a battery for more than a few seconds before its power was seriously impaired. The zinc and acid in these batteries were rather quickly consumed and were expensive; also, the acid caused much damage when it was spilled.

THE DANIELL CELL

Despite all of these objectionable characteristics, there was no substantial change in batteries until 1836, when Professor J. F. Daniell devised the battery that bears his name. This battery still had electrodes of copper and zinc, but the solution, now called electrolyte (Faraday's term), was copper sulfate and zinc sulfate. The zinc electrode was in a porous cup, in a zinc sulfate solution, and the copper electrode in cylindrical form surrounding the cup was immersed in a copper sulfate solution, all of which was contained in a cylindrical glass jar. This battery did not polarize and used neutral salts that were only mildly corrosive, and it was not expensive. It delivered about 1.1 volts as we measure electromotive force today. A modification of the Daniell cell, known as the gravity, crowfoot, or bluestone battery, was widely used on American telegraph lines for many years. In this battery, the two fluids were separated from each other by the difference in their specific gravities. The zinc sulfate, being lighter, floated above the copper sulphate. In its usual form the zinc, or negative, element was a heavy casting with radiating fingers (hence the name crowfoot) which was clamped over the top of the battery jar by a lug opposite the fingers. The copper element was made up of three or five copper leaves, held together by a copper rivet in the center, with the outer leaves bent outward in arcs to form a somewhat star-shaped figure. The copper element was laid on the bottom, immersed in the copper sulfate solution, and was connected to the outside by means of an insulated copper wire that ran up the side of the jar and was bent over the edge.

In another modification of the Daniell cell, devised by J. H. Meidinger, a spherical glass flask with a long neck was filled with copper sulfate solution, with the neck immersed in the part of the Daniell cell outside the porous cup. When a battery of any of these types was in operation, copper from the copper sulfate solution was deposited on the copper electrode, while at the same time the zinc element was converted into zinc sulfate by chemical action of the free SO_4 ions, released by the copper sulfate, attacking the zinc electrode. The effect was to increase the quantity of zinc sulfate and diminish the quantity of copper sulfate; and therefore, as the battery operated, some of the zinc sulfate solution had to be removed from time to time, and the copper sulfate replenished. In the Meidinger battery this was accomplished by means of the flask, and in the gravity battery copper sulfate crystals were placed in the bottom, zinc sulfate drawn off, and the water replenished.

THE GROVE CELL

Sir W. R. Grove, in 1839, devised a cell in which a zinc electrode was placed in a porous cup in a dilute sulfuric acid solution, and outside the cup was a platinum plate in concentrated nitric acid. The Grove battery had a high electromotive force, about 1.9 volts, and it did not polarize, but it was objectionable because it gave off disagreeable brown fumes produced by nitric acid, and the strong acids were highly corrosive. Furthermore the platinum electrode was very expensive.

Bunsen altered the Grove cell somewhat in 1842 by substituting a carbon rod for the platinum element. This change reduced the cost, but the other objections to the Grove battery still remained. The Bunsen cell was, nevertheless, used rather extensively. Its voltage was the same as the Grove battery, about 1.9.

THE LECLANCHÉ CELL

The most widely used batteries were those invented by Georges Leclanché, which were known by that name. The elements in the Leclanché battery were zinc and carbon in an electrolyte of ammonium chloride (sal ammoniac). The earlier forms, called wet batteries, were made with a large carbon cylinder, in the center of which was a zinc rod separated from the carbon by an insulator. These batteries polarized rather quickly, but this fault was partially overcome by making the carbon electrode large and porous so as to absorb and distribute the hydrogen that appeared on the positive electrode.

A later form of the Leclanché cell was the dry battery, which was widely used in telephone work and for many other purposes, especially flashlights. In its usual form, the dry cell consists of a zinc container, which is also the negative electrode, the inside of which is lined with paper and filled with a mixture of granulated carbon and manganese dioxide saturated with ammonium chloride, with a carbon electrode in the center. More recent dry cells have a steel jacket on the outside to prevent damage by efflorescence. Polarization is minimized by the manganese dioxide, which gives up oxygen to form water with the hydrogen.

OTHER BATTERIES

There were numerous other forms of batteries which were used at various times. Among these were the Fuller, Krüger, Lalande-Chaperon, Partz, Grenet, bichromate, and Smee. The Edison-Lalande battery is somewhat similar to the Lalande-Chaperon battery except that in the Edison-Lalande battery the positive electrode is copper oxide instead of iron, but the electrolyte in both cases is a solution of caustic potash. The Edison-Lalande battery was rather widely used in the United States for railway signal work, because, among other reasons, the battery did not freeze.

Two standard cells have been introduced for use in the laboratory. One of these is the Weston cell which has electrodes of cadmium and mercury and an electrolyte of cadmium sulfate. Its voltage is 1.0183 at 20° Celsius. The other is the Clark cell with electrodes of zinc and mercury and an electrolyte of zinc sulfate. Its voltage is 1.4328 at 15° Celsius.

In recent years several types of dry cells have been produced in quantities because of their special uses. They have such advantages as greater discharge capacities and longer life; in some cases they are rechargeable. These batteries are not necessarily new.

There is a mercury battery that has a potassium hydroxide electrolyte. The nickel-cadmium battery is rechargeable. Another rechargeable battery is the manganese-zinc battery with a potassium-hydroxide electrolyte. If completely discharged this battery is ruined. Otherwise it may be charged up to about fifty times. The silver chloride battery dates back to 1860 and is still used for special purposes.

STORAGE BATTERIES

The principle of the secondary, or storage, battery was discovered early, but for many years was of only scientific interest because there was nothing to be gained by charging one battery with another; the situation changed greatly when the dynamo appeared. Gaston Planté, who had been investigating polarization, made the very important discovery in 1860 that lead plates immersed in a sulfuric acid solution and charged constituted an excellent secondary battery. When such a battery was charged, a thin layer of lead dioxide was formed on the positive plate, while the negative plate was converted into sponge lead. Planté's batteries were formed by repeated charging and discharging, a process that was slow and expensive. In his earlier batteries he rolled together two sheets of lead, separated by heavy linen, gutta-percha, or rubber. Later he constructed batteries with multiple flat plates in which the positive and negative elements were interleaved, as in a modern storage battery.

Inspired by Planté's successes, others began to experiment with secondary batteries, notably Camille Faure, also a Frenchman, who in 1881, patented a battery with lead plates to which had been applied a coating of red lead. The red lead was more easily converted to lead dioxide and sponge lead than the pure lead in Planté's battery. The Faure battery also had more active material and hence greater capacity.

A further improvement was made by Sellon and by Volckmar, who prepared the plates with grooved or perforated faces, and the depressions were then filled with red lead paste before the forming process. Batteries of this type, with further improvements, were built by Swan in England and Brush in the United States. Another improvement that followed shortly was the use of cast grids containing a small quantity of antimony.

ELECTROSTATIC INDUCTION MACHINES

Electrical currents of considerable magnitude were now available from batteries, magnetos, and dynamos, but there was still great interest in the older electrostatic machines. There were many of the old frictional machines in use for experimental purposes in laboratories, but Volta's electrophorus and Bennet's doubler had shown that an original small charge could be increased many times by manipulation. Holtz and Toepler in Germany and Wimshurst in England devised new electrical machines based on the induction principle, so arranged that after a few turns a small initial charge was increased to a very large one. In these machines, as well as in an earlier one by C. F. Varley, sectors of metal foil were mounted on glass disks, which in turn were mounted on shafts so that they could be revolved. Charges were built up by induction from disks on a neighboring plate. In the Holtz and Toepler machine, the second disk was stationary, whereas in the Wimshurst machine the plates revolved in opposite directions.

These new electrostatic machines were built in great numbers for experimental purposes and also for the medical profession, to be used for the treatment of disease. Volumes were written on the subject of electrotherapeutics, and electrodes of all conceivable shapes were designed to reach the various organs or tissues to be treated. No progressive or prosperous physician's office was without a Holtz or Wimshurst machine and a multitude of accessories. Some of these machines were of great size, and they performed miracles, if the reports of the period could be believed, but miracles notwithstanding the machines gradually disappeared from the physicians' offices.

We may doubt the testimonials regarding the alleviation of human suffering, but these machines did contribute in a very substantial way to the increase of knowledge and afforded much entertainment. One of the most beautiful experiments that could be performed with such a machine was with Geissler tubes. Heinrich Geissler was a glassblower at the university at Bonn, Germany. His tubes were made in many fantastic shapes; some were evacuated, while others contained traces of various gases or mercury. Many of the glass envelopes contained metallic salts that fluoresced when excited by the electrical machine. In some respects, Geissler tubes anticipated modern tube or fluorescent lighting.

During the early part of this period, the most important development was the introduction of the electromagnetic telegraph, but this is a story in itself and will be told later.

The electric bell was the first electrical device to be used extensively in homes. The inventor of the electric bell is not definitely known, but it may have been Miraud of Rouen, France, or his compatriot Leclanché, the inventor of the sal-ammoniac battery.

Before the large-scale commercial use of electricity for lighting, there were electrical concerns manufacturing telegraph instruments, insulated wire, cable, batteries, insulators, bells, and many telegraph accessories. These concerns were often instrumental in the development of new apparatus such as annunciators, burglar alarms, fire alarms, and railway signals. As the scope of the industry widened, several of these firms took over the manufacture of dynamos, lighting equipment, switches, sockets, fixtures, fuses, and a hundred other items. Some of these companies became the nuclei of the great electrical manufacturing concerns of today.

7

Electrical Instruments, Laws, and Definitions of Units

Before electricity could become an exact science, capable of mathematical analysis, it was essential that its properties be defined in terms of measurable units. Certain concepts regarding electrical power, pressures, quantities, rates of flow, and those relating to circuits, such as resistance, self-induction, capacity, and still others relating to dielectrics and chemical action had already been well established before the units of measurement were defined. Similar ideas regarding magnetism and magnetic circuits existed.

TANGENT GALVANOMETER

A great many instruments of various kinds have already been mentioned in previous chapters, some of which were mere indicators, and others provided accurate quantitative measurements. A great improvement in the Schweigger and Poggendorff multiplier was made by Pouillet in 1837, at which time he invented the tangent galvanometer. This instrument had a relatively large coil with a diameter of 30 centimeters or more, at the center of which was a short magnetic needle to which a pointer was attached at right angles that moved over a scale divided into degrees. When no current flowed through the coil, the needle came to rest in the magnetic meridian, but when a current was sent through the coil, with the plane of the coil in the magnetic meridian, the needle took a position which was in the direction of the resultant of the forces produced by the earth's magnetic field and the field produced by the current flowing through the coil. By reading the deflection of the needle in degrees, the current could be calculated by multiplying

Figure 7.1 Pouillet's Tangent Galvanometer *(Courtesy Burndy Library)*

the constant for the particular instrument by the tangent of the angle of deflection. Weber greatly improved this instrument in 1846 by attaching a small mirror to the needle, in which the reflections of a scale could be read through a telescope. In the same year Weber also built a moving coil galvanometer, as opposed to the stationary coil in the tangent galvanometer. This instrument depended only on the interaction between the field produced by the coil and the earth's field.

D'ARSONVAL GALVANOMETER

The most reliable of all galvanometers was that invented by Deprez and d'Arsonval in 1862, known as the d'Arsonval galvanometer. In this instrument a small moving coil was suspended between the poles of a permanent magnet. The magnet was provided with concave pole faces. A cylindrical iron core was mounted between the pole faces, leaving a gap on either side sufficient for the coil to swing without touching. Originally the coil was suspended on torsion wires that were also lead-in wires for the coil. Later the coil was mounted on jeweled bearings with hairsprings used as a restraining force and also used as conductors for the current.

WHEATSTONE BRIDGE

Wheatstone's bridge, as it is commonly known, although according to Silvanus P. Thompson, it was invented by Christie, came into use about 1843. It was the first accurate device for comparing resistances. In it an unknown resistance could be inserted into a divided circuit that included three known resistances, a battery, and a galvanometer. When the known resistances, one of which was variable, were properly balanced so that the galvanometer reading was zero, the unknown resistance could be calculated by a simple proportion.

ELECTRICAL AND MAGNETIC LAWS

The advent of accurate measuring instruments made it possible to verify and extend electrical and magnetic laws that in many cases had been proposed on the basis of partial information. Some of these laws have already been mentioned, such as the inverse square laws for electric and magnetic fields by Coulomb, the law governing the force between a conductor carrying a current and a magnetic pole by Ampère, and Ohm's law, which was announced in a paper in 1827, entitled "Die Galvanische Kette."

Before units of measurement for electricity and magnetism could be established, it was necessary to define certain fundamental relationships by laws expressed in the form of equations. It is regrettable that we do not know the authors or the years for some of these laws, particularly the magnetic laws, but it is probable that many if not all of them were the work of Gauss. Between the years 1820 and 1862 electrical and magnetic laws, measurements, and units were firmly established through the efforts of a long list of workers including Ampère, Coulomb, Biot, Savart, Faraday, Henry, Ohm, Poisson, Pouillet, Gauss, Weber, Lenz, Riemann, Kirchhoff, Neumann, Kohlrausch, Green, Joule, Lord Kelvin, von Helmholtz, and Maxwell.

The accomplishments of these men were greatly influenced by the work of Newton as set forth in his *Principia*. Newton was the first scientist to apply mathematics to such general natural laws. Newton had established the inverse square law for gravitational fields, although Hooke had claimed credit for the principle. Coulomb was undoubtedly aided by this law for gravitational fields in

Table 7.1. Fundamental Equations

Electric Equation	Author	Year	Magnetic Equation	Author	Year
$f = k\, qq'/r^2$	Coulomb	1785	$f = k\, mm'/r^2$	Coulomb	1785
$H = 2i/r$	Biot and Savart	1820	$H = 4\pi ni/l$	–	–
$f = im \cdot ds/r^2$	Ampère	1822	$H = 2\pi ni/r$	–	–
$q = it$	–	–	$H = f/m$	–	–
$I = V/R$	Ohm	1827	$\phi = BA$	–	–
$V = d\phi/dt$	Faraday	1831	$\mu = B/H$	–	–
Heat $= ki^2Rt$	Joule	1843			

formulating his law for electric and magnetic fields, which he confirmed by means of his torsion balance.

Table 7.1 gives the fundamental electric and magnetic equations upon which the definitions of units were based.

ELECTRIC AND MAGNETIC UNITS

Karl Friedrich Gauss (1777–1855) and Wilhelm Weber (1804–1891) were the principal authors of the electric and magnetic units. Gauss developed a system of magnetic units in 1832, using the millimeter, milligram, and second as the units of length, mass, and time. In 1839 Gauss defined electric potential at a given point in the electrostatic system as the work required to move a unit charge from infinity to that point. Beginning in 1840, Weber developed the definitions of the electric units in the electromagnetic system. He also, as did Gauss, used the millimeter, milligram, second system. Other scientists generally preferred the centimeter, gram, second system, or, as it is generally known, the cgs system.

For some years after Gauss and Weber there were no widely accepted units for electricity and magnetism. In 1861 the Committee on Electrical Standards of the British Association for the Advancement of Science, under the leadership of William Thomson (later Lord Kelvin), was given the task of formulating definitions that would be consistent with one another and would be reproducible. They recognized the work already accomplished by Gauss and Weber and proceeded from that point. In the committee's report of 1864 it established values for the practical units of electromotive force and resistance, based upon the meter, gram, second system. This committee established the practical unit of electromotive force, which it called the volt, at 10^8 times the absolute electromagnetic unit. The practical unit of resistance, which this committee called the ohm, was established at 10^9 times the emu absolute unit. From these definitions of the volt and ohm the practical unit of current would become $^1/_{10}$ the emu absolute unit.

A later committee of the British Association recommended in 1873 the use of the cgs system instead of the mgs system. This committee also recommended the use of the terms erg and dyne for the absolute units of work and force. These recommendations met with general approval throughout the scientific world.

The first International Electrical Congress met in Paris in 1881 and approved the cgs system as the basis for the definitions of the electric and magnetic units. The Paris International Congress confirmed the definitions established by the British Association in 1864 and 1873. It also defined and named the coulomb and farad.

Neither the British Association nor the International Congress gave any consideration to the esu absolute system of units. Gauss and Weber had developed both the emu and

Figure 7.2 Lord Kelvin *(From Burndy Library)*

esu absolute electric and magnetic units. Weber noted the numerical ratio between the two systems in 1846. In 1856 Weber and Kohlrausch found that the dimensions of this ratio constituted a velocity which equaled the velocity of light, within the limits of experimental error. The velocity of light was known approximately at that time from the observations of Roemer in 1676 on the eclipses of Jupiter's satellites. The first accurate, terrestrial measurements were made by Fizeau in 1849.

The fourth International Electrical Congress met at the World's Fair in Chicago in 1893. It confirmed the previous definitions of the absolute and practical units and defined a prototype of the standard ohm as the resistance of a column of mercury 106.3 centimeters in length, and 1 square millimeter in cross section at a temperature of 0° Celsius. The prototype of the volt was given in terms of the voltage of the Clark standard cell whose voltage was given as 1.434 volts at 0°Celsius.

There were periodic meetings of the International Electrical Congress at which refinements in the definitions were discussed but in general the original absolute and practical units remained unchanged. Suggestions were made from time to time for modifications in the definitions of the units. In 1882 Heaviside suggested changing the definition of the unit magnetic pole so as to eliminate the factor 4π that occurs in certain of the fundamental equations. Föppl in 1894 proposed changes in the coefficients of the fundamental magnetic and electric units for the sake of greater symmetry. Lorentz called this the Gaussian system. Giorgi in 1901 proposed a combination of these suggestions with the mks (meter, kilogram, second) system. These and other proposals will be discussed later.

Table 7.2 shows the ratios of the principal practical electric units to the emu and esu absolute units. Table 7.3 is a corresponding table for magnetic units.

It is somewhat ambiguous to speak of practical magnetic units since the emu units are used in most practical applications.

The subject of electric and magnetic units is rather confusing so that in reading a textbook one must first discover which system of units the author is using. Systems other

Table 7.2. Practical Electrical Units

Kind of Unit	Name of Unit	Symbol	Ratios of Practical Units to emu	esu
Potential	Volt	V	10^8	$1/(3 \times 10^2)$
Current	Ampere	I	10^{-1}	3×10^9
Resistance	Ohm	R	10^9	$1/(9 \times 10^{11})$
Quantity	Coulomb	Q	10^{-1}	3×10^9
Capacity	Farad	C	10^{-9}	9×10^{11}
Self-Induction	Henry	L	10^9	$1/(9 \times 10^{11})$

Table 7.3. Practical Magnetic Units

Kind of Unit	Name of Unit	Symbol	Ratios of Practical Units to emu	esu
M.M.F.	Gilbert	$\mathfrak{F}$	1	3×10^{10}
Flux	Maxwell	ϕ	1	$1/(3 \times 10^{10})$
Reluctance	$\frac{\text{Gilbert}}{\text{Maxwell}}$	$\mathfrak{R}$	1	9×10^{20}
Intensity	Oersted	H	1	3×10^{10}
Permeability	$\frac{\text{Gauss}}{\text{Oersted}}$	μ	1	$1/(9 \times 10^{20})$
Induction	Gauss	B	1	$1/(3 \times 10^{10})$

than the practical, emu, or esu which have been introduced are the following: mks (meter-kilogram-second), mksa (meter-kilogram-second-ampere), Giorgi (Same as mksa), Heaviside-Lorentz (rationalized system that suppresses the factor, 4π), and Gaussian (electric units numerically equal to esu and magnetic units equal to emu).

The IEC (International Electrotechnical Commission), which was organized in 1904, is now an arm of UNESCO in the field of electrical engineering. It and its committees are now responsible for the definitions of electric and magnetic units. In 1935 the EMMU committee of the IEC adopted the mks system, and in 1950 the U. S. Congress passed a law fixing the electric units for the United States as defined in the mksa system.

The subject of electric and magnetic units has now become so involved that it would be impossible to discuss the matter fully in this history. Volumes have been written on this subject, but the best and most concise review of the present state of the matter is U. S. Bureau of Standards Monograph 56 written by the late Francis B. Silsbee, published in 1962, and reprinted in 1963.

8

The Electric Telegraph

The need for rapid communication between distant points has been felt from the beginnings of civilization. Many early methods were devised, some of which operated with a fair degree of success and are still in use. Among these devices are semaphores, smoke signals, drums, the firing of cannon, heliographs, blinker lights, bells, and others. In the time of Pierre de Maricourt it was proposed to use compass needles for signaling, on the theory that the movement of one compass needle would cause a similar movement in another needle, even at a great distance. Magnetic needles did in fact become part of certain telegraphic devices at a later date.

After Stephen Gray's discovery of the transmission of electricity over conducting threads and the experiments that followed, in which Leyden jars were discharged over great lengths of wire, the thoughts of many turned to the use of electricity for communication. Gray's experiments were made in 1729, and in 1746 Winckler set up an electric telegraph in which there were as many wires as there were characters to be transmitted.

Scots Magazine published in 1753 an anonymous letter, later ascribed to Charles Morrison, which proposed using many wires each of which would terminate in a metal ball corresponding with a letter of the alphabet. Near each ball was to be hung a strip of paper, properly labeled, which would be attracted when a charge was sent over the line. He also suggested the use of glass insulators for supporting the wires.

Lomond proposed in 1787 using a single wire with a pith ball electroscope at the receiving end and, presumably, using some sort of code. Reizen built a multiple-wire telegraph in 1794 in which sparks at the receiving end were used to identify the characters transmitted. Sömmering built a multiple-wire telegraph at Munich in 1809, using Volta's recently discovered batteries and Carlisle and Nicholson's discovery of the decomposition of water. Each of the wires in Sömmering's telegraph terminated in a cup of acidulated water, so that when the circuit was closed, through the battery, bubbles of gas were liberated in the appropriate cell.

EARLY ELECTROMAGNETIC TELEGRAPHS

After Oersted's discovery of electromagnetism and Schweigger's invention of his galvanometer or multiplier, both Ampère and Laplace proposed electromagnetic telegraphs employing coils and magnetic needles at the receiving end, equal in number to the characters to be transmitted, with a corresponding number of wires. Harrison Gray Dyar built a telegraph line on Long Island, New York, which operated during the years 1828 and 1829, in which messages were recorded on a moving strip of paper by electrochemical means.

Joseph Henry's line, built in 1830, has already been mentioned. On this line he used the equivalent of a telegraph sounder. Weber and Gauss built a telegraph line at Göttingen in 1833, which was nearly a mile in length, in which the receiving device was a galvanometer. In the same year Baron Schilling constructed a telegraph line at St. Petersburg, using five needles in his receiving apparatus.

Wheatstone's telegraph of 1837 employed a single needle and was the first to be used commercially. A considerable amount of telegraph business developed in England during the ensuing years. In 1837 Steinheil constructed a telegraph line at Munich, more than 12 kilometers in length, using a receiving apparatus that recorded the message on a paper tape by means of dots. Steinheil also discovered in 1838 that it was possible to use the earth for one side of the circuit.

SAMUEL F. B. MORSE

Many others were giving serious thought to the problem of the electric telegraph, among whom was Samuel F. B. Morse (1791-1872), who had acquired a limited scientific knowledge during his undergraduate days at Yale University. At this time, however, Morse was far more interested in painting than in science, and to this end he spent the years 1811 to 1815 in England studying art. Upon his return to the United States he became a successful portrait painter, and in 1825 was one of the founders of the National Academy of Design. His interest in science was revived in 1827 when he studied electromagnetism under J. F. Dana of Columbia College. He was still, however, interested primarily in painting and in 1829 returned to Europe to study the great works of art in the galleries of England, France, and Italy.

On his return voyage to the United States aboard the packet ship *Sully*, he met Dr. Charles S. Jackson, who described to Morse a number of electrical experiments he had witnessed a short time before in Paris. These conversations with Dr. Jackson renewed Morse's interest in electromagnetism, whereupon he confided to Dr. Jackson his latent ideas about an electric telegraph. He soon became completely absorbed in the matter to the extent that he could think of nothing else. He worked out in considerable detail the design of an electromagnetic telegraph recorder, a mechanical sending device, a code of dots and dashes for numbers, and perhaps also a relay.

Figure 8.1 Samuel Morse *(From Burndy Library)*

Before the *Sully* reached port, Morse had made rather complete sketches and notes on the recording telegraph he intended to construct. The *Sully* docked at the wharf at the foot of Rector Street in New York on November 15, 1832. As he disembarked, Morse was met by his two brothers, Sidney E. and Richard C. Morse. Because he was almost without funds at the time, his brother Richard invited him to live at his house until such time as he might again become self-supporting. Morse's brothers were the publishers of a newspaper, the *New York Observer*, and were apparently comfortably well-off.

The telegraph that Morse had in mind at this time was designed to send and receive by mechanical means. In his sending apparatus he planned to use a form of type consisting of dots and dashes. He was given a room at his brother's house where he began immediately to work on his telegraph. Here he cast type and in the process damaged the furniture and carpets. His brother then gave him the use of a room of the fifth floor of the newspaper building, at the corner of Nassau and Beekman Streets. This room became his shop, studio, kitchen, living room, and bedroom. Here he worked, ate, slept, and lived in the direst poverty. It is not clear how his three children were cared for. They had been living with relatives since the death of their mother some years earlier.

Morse labored on his telegraph in these quarters for some time and then moved to Greenwich Lane. Here he eked out a bare living by painting an occasional portrait and by giving lessons, but his primary concern was with his telegraph, which began to take shape in a very crude form. His transmitting instrument was made with a wood frame on which was mounted a pair of wooden rollers, one of which was provided with a crank. A belt made of carpet binding passed over the rollers and carried the portrule, a sort of typesetter's stick, in which the lead type was placed. A lever was mounted above the rollers on a pivot near its center, one end of which was provided with a rounded tooth that engaged the projections on the type as the portrule was carried forward. At the other end of the lever, a wire bent into the form of an arc was attached with its ends dipping into two mercury cups.

The receiving apparatus of this first Morse telegraph was also built on a wood frame. It carried three wooden rollers, over which passed a strip of paper. One roller was driven by clockwork that moved the paper strip lengthwise. Over the middle roller was a pendulum whose swing was limited by two stops. On its lower end, it carried a pencil that was made to move crosswise of the paper strip under the influence of an electromagnet. The circuit passed from one pole of the battery through the mercury cups, to the line, to the electromagnet and returned to the battery over the other side of the line. Impulses corresponding to the dots and dashes, set up in type on the portrule, were sent through the circuit, causing the pendulum to swing and causing the pencil at the lower end of the pendulum to draw an irregular sawtooth line. The mark on the paper could then be translated into dots and dashes and, by means of the code, into numbers. Morse's code at this time was not alphabetical but contained only numbers, which could be translated into words by means of a dictionary.

In 1835 Morse was appointed professor of the Literature of the Arts of Design at New York City University, and was assigned to quarters in the university building on Washington Square. He moved his belongings to the new quarters, including, of course, his telegraph.

Progress up to this time had been painfully slow, but

within a short time, perhaps with the help of Professor Page, he succeeded in getting his telegraph to operate, and late in 1835 was able to exhibit his invention to a group of friends. Professor Page was Charles Grafton Page, a native of Salem, Massachusetts, and a physician turned physicist and inventor. He contributed greatly to the development of the induction coil. He built several motors of the intermittent impulse type, one of which he used to operate a small locomotive. He was also the author of several books on induction.

In 1836 Morse continued his work on the telegraph in order to improve its operation. He also experimented with recording by chemical means, a method that had previously been carried out by Dyar, and later by Bain, but Morse decided against it. Also during this period, Morse gave further thought to the matter of relays and in 1835 or 1836 designed such a device.

DEMONSTRATION OF THE FIRST MORSE TELEGRAPH

On Saturday, September 2, 1837, Morse exhibited his telegraph, operating over 1700 feet of wire, to a group which included Professor Daubeny of Oxford University, Professor L. D. Gale of New York City University, Professor Torrey, and Mr. Alfred Vail, who was a recent graduate of the university and whose family owned the Speedwell Iron Works of Speedwell, New Jersey. The Vail business included a brass and iron foundry and a machine shop.

PARTNERSHIP WITH ALFRED VAIL

Alfred Vail became greatly interested in Morse's telegraph, and returned the following week to discuss the invention more fully with Morse. He was convinced that the telegraph was practical and entered into a partnership agreement with Morse under which the Speedwell Iron Works would undertake the construction of an improved instrument. This arrangement was vital to the success of the telegraph and resulted in the production of much better and more reliable instruments.

U. S. GOVERNMENT INTERESTED IN TELEGRAPH

On March 10, 1837, the secretary of the treasury, Levi Woodbury issued a circular letter to collectors of customs, commanders of revenue cutters, and other interested persons regarding a resolution passed by the House of Representatives, which requested information on a system of telegraphs for the government of the United States. Morse, who had received a copy of the circular letter, sent a detailed description of his telegraph and its mode of operation to the secretary, dated September 27, 1837, and on November 28, 1837 sent a second letter giving the secretary the results of the demonstration at New York City University in September.

The Speedwell Iron Works was in the process of making an improved Morse telegraph, which was completed late in 1837 and was demonstrated in a room at the Speedwell shop on January 6, 1838, over a 3-mile circuit of insulated copper wire. It is not clear whether the new instrument used a fountain pen or merely a stylus to indent the paper, but both methods were used in Morse telegraph instruments. The new receiving instrument made four separate lines of dots and dashes on the paper, probably on the theory that at least one would be legible. The code in use was still the early one, using only numbers that were translated into words by means of a specially prepared dictionary.

DEMONSTRATIONS OF THE IMPROVED MORSE TELEGRAPH

The telegraph was now ready for a public demonstration, and such a demonstration was held before a large gathering in the Geological Cabinet of the University of the City of New York on January 24, 1838. Ten miles of wire were used in this test, and for the first time what is now known as the Morse code was employed. In other words, messages were composed of characters representing letters and not numbers. There is some evidence to indicate that Vail rather than Morse may have been responsible for this new telegraphic alphabet, although Morse was undoubtedly consulted in the matter. Whether or not the portrule for mechanical sending was still in use at this time cannot be determined definitely. Vail had improved the Morse telegraph greatly and had transformed it from a crude working model to an instrument that in appearance and performance was of commercial quality.

A more important demonstration was given on February 8, 1838, before the Franklin Institute of Philadelphia, perhaps the foremost scientific body in the United States at the time. Its Science and Arts Committee was greatly impressed by what it saw.

From Philadelphia, Morse took his telegraph to Washington, where on February 21, 1838, President Van Buren and his cabinet witnessed the transmission of telegraphic messages over 10 miles of wire. Preceding this demonstration, Morse had written a letter to F. O. J. Smith, chairman of the Committee on Commerce of the House of Representatives, in which he proposed building an underground line 100 miles long and offered the telegraph for the sole use of the government because he foresaw a possible misuse of the facility in private hands. After the demonstration, Morse, in a second letter, reduced the suggested distance to 50 miles and requested an appropriation of $26,000, excluding contingencies, to finance the construction of an experimental line.

Following Morse's letter to Smith, the Commerce Committee reported to the House of Representatives on the demonstration and recommended an appropriation of $30,000. Mr. Smith then decided that he would like to join the telegraph venture and offered to resign his position in Congress. A partnership was agreed upon between Morse and his associates in which sixteen shares were divided as follows: Morse nine shares, Smith four shares, Vail two shares, and Professor Gale one share.

PATENT APPLICATIONS

On April 7, 1838, Morse applied for a United States patent. The partners believed that foreign rights to the invention should be protected by patents as well as the rights at home, and accordingly Morse and Smith sailed for England on May 16, 1838. After spending seven weeks of frustrating effort with British officials, they were refused a patent, partly on the grounds that a telegraph had already been patented by Wheatstone and Cooke, and partly because the claim was made that the details of Morse's telegraph had already been published. Both arguments appear to have been insincere, and the truth was probably that the British government did not wish to have an American firm invade the British telegraph business. Although Morse was greatly disappointed in his efforts to obtain a British patent, he found pleasure in being able to attend the coronation of England's great Queen Victoria as the guest of the American minister, Andrew Stevenson.

Morse and Smith left England on July 26, 1838, on their way to Paris. In Paris Morse exhibited his telegraph to a distinguished gathering which included Arago, Baron Humboldt, Gay-Lussac, and others. In describing the apparatus exhibited in Paris, Morse again referred to the portrule, indicating that he had still not given up the idea of mechanical sending.

Morse remained in Paris until March of the following year, during which time he exhibited his apparatus to many groups and individuals. There was at this time much political unrest in France involving King Louis Philippe, and as a result the patent application was delayed. He did, however, receive a patent eventually, but it had little value because of certain oddities in the French patent law, which provided that an actual installation must be made within two years. Since the telegraph was to be a government monopoly, and since the French government was not disposed to make the necessary expenditures, nothing came of the patent.

While in Paris, Morse made application for a Russian patent through the Russian minister, Baron Meyendorf. Shortly thereafter, he returned to England and held further demonstrations before members of the House of Lords, after which he sailed back to the United States aboard the steamer *Great Western,* which arrived in New York on April 15, 1839.

Because Morse and his partners lacked funds and because Congress had not acted on the bill providing money for building an experimental line, there was little progress during the next several years. After more than two years a United States patent was issued to Morse on his telegraph on June 20, 1840. The delay was not so much the fault of the patent office as it was Morse's failure to supply all of the necessary information. There was no action by Congress on the appropriation bill during the sessions of 1840, 1841, and 1842.

SUBMARINE CABLE

During this rather dismal period for Morse and his partners there was one interesting new development. Morse somehow got together enough money to have two miles of submarine cable manufactured, which he laid between the Battery and Governor's Island in New York harbor in the summer of 1842. A test of this cable was made on October 19, 1842, during which four characters were received when transmission was suddenly interrupted. A ship's anchor had fouled the cable, and when it was drawn to the surface the ship's crew cut it off because they had no idea as to its purpose. This cable contained a single strand of copper wire, insulated with rubber and protected with spiral strands of tarred hemp.

CONGRESS APPROPRIATES $30,000 FOR AN EXPERIMENTAL LINE

On December 30, 1842, Charles G. Ferris, a member of the Commerce Committee of the House of Representatives, submitted a report on Morse's telegraph together with a bill calling for the appropriation of $30,000 for the construction of a telegraph line by Samuel F. B. Morse and his associates. This bill passed the House on February 23, 1843, by a vote of eighty-nine to eighty-three. The Senate was preoccupied with many matters, but finally on the last day of the session, March 3, 1843, just before midnight, the bill was approved by the Senate without opposition. Morse had been sitting in the gallery all day and into the evening, hoping that the telegraph bill would be called up. When he was told that there was no likelihood that the bill would be considered, he left the Senate Chamber with a heavy heart. He was informed of its passage early the next morning by

Miss Ellsworth, the daughter of the commissioner of patents, who had become a friend of Morse.

CONSTRUCTION OF THE LINE

Morse and his partners were jubilant and began immediately to make preparations for the construction of a telegraph line between Washington and Baltimore, to be laid underground in lead pipe. Vail began the manufacture of new telegraph instruments. Professor L. D. Gale was put in charge of construction. Professor J. C. Fisher supervised the manufacture of the insulated wire, and Ezra Cornell, who had invented a machine for laying pipes underground, was put in charge of laying the lead pipe.

It became evident after some miles of pipe had been laid that the underground system would not work. Apparently the difficulty lay in the fact that the joints in the pipe were being made with the insulated wire in place, and the heat used in splicing charred the insulating material and caused it to break down. The lead pipe was probably in short lengths, and no junction boxes were used.

Two-thirds of the appropriation had been spent when the underground project was abandoned, but some of the materials purchased could be salvaged. It was then decided to erect poles along the Baltimore and Ohio Railroad right-of-way. Morse referred to the poles as spars, and they were described as 30 feet high, probably the overall length. They were placed at intervals of 200 to 300 feet, a distance which was too great for the small-sized copper wire used. The wire was probably no larger than No. 14, and with these long spans there was difficulty later with the breaking of wires and of wires swinging together in high winds. Sufficient wire had been purchased to span four times the distance between Baltimore and Washington, but it is unlikely that all of the wire was used. The wire was attached to insulators, but there is no description of them. Later it was the practice on American telegraph lines to use glass insulators, whereas on European lines porcelain was used.

In earlier experiments with the telegraph, Morse had used Cruikshank batteries, with which he experienced considerable difficulty because of polarization, but on this project he turned to the Grove battery, which was expensive but did not polarize. On his latest European trip Morse had learned of the Daniell cell, which he admired greatly but which for some unknown reason he did not use on this line. Later a modification of the Daniell cell, known as the gravity cell, became the American standard for telegraph use.

As the line progressed, constant testing was carried on from various points along the line to the terminal in Washington. When the line reached the railway junction near Baltimore, a temporary station was established from which messages could be exchanged with Washington. Work had then been in progress for about a year, and it was decided that the line must be in operation by May 1, 1844, at which time the National Whig Convention was to meet in Baltimore.

When Henry Clay was nominated for president by the Convention, the news was carried by messenger to the terminus of the then-existing line and sent to Washington by telegraph. Similarly, the nomination of Theodore Frelinghuysen for vice-president was sent to Washington. Shortly thereafter the telegraph circuit was completed to Mount Clare depot in Baltimore and to the Supreme Court building in Washington.

"WHAT HATH GOD WROUGHT!"

On May 24, 1844, the first official test of the line was to be made. On the morning of that day Morse, his friends, and other observers gathered about the instrument in the Supreme Court building in Washington while Alfred Vail was at the instrument in Baltimore. Morse had agreed to allow Miss Ellsworth to select the message to be sent, as an acknowledgment of his gratitude for her service in bringing him the news of the passage by the Senate of the appropriation bill. Miss Ellsworth chose a short passage from the Book of Numbers in the Bible, "What hath God wrought!" The message was sent by Morse and received by Vail in Baltimore, who promptly repeated the message back to Washington. Morse was congratulated by his friends and soon by official Washington as the news of his success spread.

It is impossible to say precisely how the operation was carried out, but it seems clear that Morse had abandoned his portrule and had changed to some form of key. In describing a later test, Mr. Holmes of South Carolina, a brother of a representative in Congress, described the operation in a room beneath the Senate Chamber in these words, "saw him operate by dipping into a phial of quicksilver the end of one wire from the battery." This method of operating was very crude, but with the aid of Alfred Vail the telegraph key soon came into existence.

After the first official message on May 24, 1844, the Washington instrument was moved to the lower level of the Capitol in readiness for messages from Baltimore in connection with the Democratic National Convention, which assembled there on May 26, 1844. After nominating James K. Polk for president, the Convention nominated Silas Wright, a senator from New York for the vice-presidency. Mr. Wright was not present at the Convention but had remained in Washington. When the news of his nomination was sent to Senator Wright by telegraph, he refused to accept it. His refusal was promptly sent back to Baltimore and announced to the unbelieving delegates, who adjourned until the following morning in order to confirm the rejection, if such was indeed the case. The telegraphic message was confirmed by Senator Wright, and thereafter the messages between Baltimore and Washington were given full credence.

COMMERCIAL OPERATION OF THE TELEGRAPH

During the following year the telegraph operated without charge to its users. Expenses were covered by an appropriation by Congress of $8000, and the telegraph was placed under the supervision of the postmaster general. Alfred Vail became the operator at Washington, and H. J. Rogers at Baltimore. Beginning April 1, 1845, a charge was made of one cent for each four characters transmitted, but the revenue was insufficient to cover the cost of operation.

Morse tried to interest the federal government in the construction of a telegraph line to New York, by way of Baltimore, Wilmington, Philadelphia, and Trenton, but he was unsuccessful. He then organized the Magnetic Telegraph Company, which was incorporated in Delaware on May 15, 1845, by special act of the legislature. It was the purpose of this company to build the line to New York. Its stockholders were Morse, Gale, Vail, Ezra Cornell, and twenty-three others.

On August 6, 1845, Morse again sailed for England, and again failed in his patent application. From England he went to Holland, where the Wheatstone telegraph was in use but was now equipped with electromagnets. Morse noted that on the line between Amsterdam and Haarlem, No. 12 iron wire was being used, which had a much greater tensile strength than the smaller copper wire on his Washington-Baltimore line.

From Holland Morse went to Hamburg by steamer and then back to Paris, where he exhibited his telegraph before the French Chamber of Deputies on November 10, 1845.

CONSTRUCTION OF NEW TELEGRAPH LINES

In 1846 Morse's United States patent was reissued with improvements up to that time. Telegraph lines were now under construction in many directions, and in June 1846 the line between Washington and New York was completed. By 1847 a line was completed to St. Louis, and in 1848 one was constructed to Louisville and Nashville. The latter line was built by Henry O'Reilly and was soon involved in a patent suit brought by Morse, in which after some years of litigation, Morse was upheld by the United States Supreme Court. This was only one of many such suits that followed.

WESTERN UNION

By 1851 there were over fifty telegraph companies in the United States. The Western Union Telegraph Company was organized originally as the New York and Mississippi Valley Printing Telegraph Company on April 1, 1851, for the purpose of building a line from Buffalo to St. Louis. This company bought progressively the lines and equipment of other smaller companies until after twenty years it had unified most of the entire telegraph system of the country. Many of the smaller telegraph companies had been in serious financial difficulties and were more than willing to sell their holdings to the larger and stronger Western Union.

PRINTING TELEGRAPHS

The original New York and Mississippi Valley Telegraph Company did not use the Morse telegraph but rather the House Printing Telegraph System, which had been developed by Royal E. House of Vermont. This system, unlike the Morse recorder, printed the message in plain letters and numerals. Similar instruments using the so-called step-by-step method had been worked out by Cooke and Wheatstone in England and improved by Stroh. These latter instruments did not, however, print but merely showed the letters to the receiving operator, whereas House's instrument printed the message on paper. The House system was first used on lines between Boston and New York and between New York and Philadelphia. It continued in use on certain lines until 1860 when it was replaced by the simpler Morse telegraph with key and sounder. The sounder was first used in 1858, much to the chagrin of Morse himself, whose operators had learned to read the messages coming into his recording machine by sound.

In later years one of the stock tickers was devised on somewhat the same principle as the House printing telegraph, but there were at least half a dozen stock tickers using various schemes. The key and sounder remained the standard on both American and European telegraphs for many years, but eventually the high-speed printer was developed and replaced the key and sounder altogether.

RELAYS, DUPLEX AND MULTIPLEX SYSTEMS

Many auxiliary devices and improvements were produced as the telegraph business grew. The relay, which Morse had invented at an early stage, was greatly improved and was made very sensitive. In order to reduce the capital investment in lines, ways were sought to increase the capacity of the lines by speeding up the rate at which messages could be sent and by sending simultaneous messages over the same circuit. Gintl of Vienna was the first to invent a duplex telegraph in which a message could be sent in each direction at the same time. This telegraph came in 1853, and was followed the next year by other duplex telegraphs by Frischen, and Siemens and Halske. None of these was very successful, and it was not until 1871 that J. B. Stearns invented a successful telegraph of this type. Many others had worked on the problem including Stark, Edlund, Preece, Nedden, Farmer, Maron, and Muirhead.

In 1874 Edison invented his quadruplex telegraph, which made possible the sending of two simultaneous messages in each direction over the same circuit. The first multiplex system was invented by M. G. Farmer of Salem, Massachusetts, in 1852, using a commutation method, but this device was not used until many years later, probably because of the difficulty in maintaining synchronism at the two ends of the line. A similar device was exhibited by Meyer at the Vienna Exposition in 1873. Harmonic telegraphs using tuning forks were invented by Elisha Gray, Edison, and Bell.

RAILWAY TELEGRAPHS

It is interesting to note that the development of the telegraph and of the railroads took place at about the same time. In England the Wheatstone telegraph was used for train dispatching very early, probably before 1840. On American railways the use of the telegraph for this purpose did not occur until somewhat later, although Morse's first line was built on a railroad right-of-way and the Baltimore instrument was in a railway station. Many of the later telegraph lines were also built along railway lines, and the telegraph office was often in the railway station. The railway telegraphs were usually independent of the commercial telegraphs.

THE FIRST TRANSCONTINENTAL TELEGRAPH LINE

The New York and Mississippi Valley Telegraph Company changed its name to Western Union Telegraph Company on April 7, 1856. Its telegraph lines were spreading across the continent at an amazing rate. On October 24, 1861, the first transcontinental line was in service, an event which was of great importance to the federal government, because the Civil War was then in progress. It was almost eight years later, on May 10, 1869, that the first transcontinental railway line was completed. During the Civil War the telegraph played an important role in many campaigns, and this was the first major war in which either the telegraph or the railroad was used.

ELECTRICAL MANUFACTURING

The telegraph was the first electrical industry, and it brought with it the organization of numerous manufacturing concerns producing telegraph instruments, relays, line wire, insulated magnet wire, cable, poles, cross arms, insulators, pins, lightning arrestors, linemen's tools, and various other items. These industries grew to considerable size, and were in some cases the predecessors of companies that later produced dynamos, lighting equipment, and telephone supplies.

The Atlantic Cable

EARLY SUBMARINE CABLES

Submarine cables were not new at the time the earliest attempts at laying an Atlantic cable were made. Morse had laid a 2-mile submarine cable in New York harbor in 1842. In 1845, between Fort Lee and New York City, Ezra Cornell laid a cable 12 miles long, which contained two copper wires insulated with cotton and India rubber and protected by a lead sheath. This cable operated several months but was destroyed by ice.

In 1849 a British company undertook the laying of a telegraph cable between Dover and Calais, but this cable failed because the insulation consisted merely of tarred hemp, which could not withstand the penetration of seawater. The following year a new cable insulated with gutta-percha was manufactured in England and was laid across the channel in six or seven hours by two small steamers. This cable, however, soon failed due to mechanical stress. Still another cable was manufactured containing four wires, each insulated with gutta-percha and overlaid with hemp cords soaked in tar and tallow and protected against abrasion by an iron wire armor. This cable, laid in 1851, operated successfully.

In late May 1852, England and Ireland were connected by cable, and a year later there was a cable between Scotland and Ireland. In June 1853 a telegraph cable was laid between Oxfordness in England and The Hague in Holland, a distance of 115 miles, the longest cable so far.

NEWFOUNDLAND CABLE

As early as 1849 or 1850, F. N. Gisborne, an English engineer who had previously been in charge of the construction of telegraph lines in Nova Scotia and New Brunswick, proposed, to the Newfoundland House of Assembly, the construction of a telegraph line from St. John's in Newfoundland to the mainland. This line would consist of overland portions about 400 miles in length and an undersea portion of 85 miles. The route was to extend from St. John's to Cape Ray, from Cape Ray to Cape Breton by submarine cable, and across Cape Breton Island to the mainland of Nova Scotia, where connections could be made with existing telegraph lines.

When Gisborne surveyed the rugged terrain in Newfoundland, he decided that it would be necessary to put the line underground, and thereby also avoid the difficulties in maintaining the line under adverse weather conditions. He had been granted £500 by the Newfoundland House of Assembly for the survey and raised sufficient funds from other sources to begin construction. He had completed only 30 to 40 miles of the line when his funds were cut off and construction was halted.

It was necessary to obtain new capital from other sources, and therefore, in 1854 he went to New York where he met Matthew D. Field, a civil engineer, who had been engaged in building railroads and bridges. Field became interested and called in his brother Cyrus W. Field, who had retired the previous year from a successful business as a New York merchant and who, although only thirty-four years old, had acquired a substantial fortune.

Cyrus Field invited Gisborne to his home to discuss the matter of the Newfoundland line, but Field soon had wider visions including a submarine cable across the Atlantic. He arranged a meeting with Samuel Morse, who had long cherished the notion that communication by telegraph under the oceans was possible. In a letter to F. O. J. Smith dated

March 2, 1839, written in Paris, Morse said, "I see now that all physical obstacles which may for a while hinder, will inevitably be overcome; the problem is solved; man may instantly converse with his fellow man in any part of the world."

THE ATLANTIC CABLE

Both Field and Morse became enthusiastic over the possibility of establishing telegraphic communication with England. Field set about raising the necessary capital immediately. He organized a company of six partners called the New York, Newfoundland, and London Telegraph Company. The original partners were Cyrus Field, David Dudley Field, an attorney and brother of Cyrus, Peter Cooper, Moses Taylor, M. O. Roberts, and Chandler White. Later, after a charter had been granted by the Newfoundland House of Assembly, the names of Samuel Morse, Robert W. Lowber, Wilson G. Hunt, and John W. Brett were added to the list of partners. The new company acquired the Gisborne properties.

In the fall of 1854, Field went to England to order cable for the 85-mile Cabot Strait crossing, between Cape Ray and Cape Breton. When the Cabot Strait cable was delivered, it was loaded aboard a small sailing vessel that encountered a severe storm and foundered, with the loss of all of the cable on board. A second attempt was successful, and the necessary land lines were completed, so that by 1856 communication was established between St. John's in Newfoundland and New York.

The very difficult task of laying a cable on the little-known ocean floor was now to begin. The first step was to take soundings to determine the most feasible route. Both the British and the American governments were deeply interested in the project, and each authorized its navy to undertake soundings. For the British navy, Lieutenant Commander Joseph Dayman was in charge, and for the United States Navy, Lieutenant Berryman. Berryman, sailing east, followed the great circle route out of Trinity Bay, Newfoundland, and Dayman followed a westward route out of Valentia in Ireland. Both found the existence of a great plateau, which received the name of Telegraph Plateau. The greatest depth encountered on the plateau was 2400 fathoms, and there appeared to be no serious problems in the way of laying a submarine cable.

CABLE COMPANY IS ORGANIZED

On September 26, 1856, the Atlantic Telegraph Company was incorporated in England by Charles Bright, John Brett, and Cyrus Field, with an initial capital of £350,000. There is no further mention of Gisborne's original company, the Newfoundland Electric Telegraph Company incorporated in 1852, or the New York, Newfoundland, and London Telegraph Company organized in 1854. It may be assumed that the useful assets of these companies were transferred to the new company and that they were then dissolved. Gisborne's name does not again appear in connection with the cable companies, but apparently he remained active in the Canadian telegraph business, for in 1879 he was appointed superintendent of the Canadian government telegraph and signal service.

Field had left New York bound for England on July 19, 1856, to arrange for the manufacture of cable and to consult with his British partners regarding other steps that might be required for the cable project. With the organization of the new company, control passed from American to British hands, but this change was probably for the good of the undertaking because more British capital was available, the cable manufacturing companies were in England, and the British government was willing to subsidize the project with money and with ships. The United States Congress later also voted to render assistance.

CONTRACTS FOR THE MANUFACTURE OF CABLE

Early in 1857 the manufacture of 2500 miles of cable was begun by two British firms, Glass Elliot and Company of London and R. S. Newall and Company of Birkenhead, near Liverpool. The specifications for the cable called for a conductor composed of seven strands of No. 22 copper wire covered by three layers of gutta-percha. This portion of the cable was furnished by the Gutta-Percha Company, after which it was sent to the two cable companies. At the cable works the inner core was wrapped with hemp cord that had been treated with tar and other compounds. Next, the outer armor was applied which consisted of eighteen, seven-strand No. 22 iron wires. The outside diameter of the completed cable was a little more than 5/8 inch, and it weighed approximately 2000 pounds per mile. In addition to the submarine cable, some 25 miles of land cable for the shore ends were manufactured. This cable was nine times the weight per mile of the main cable, and was more heavily armored to protect against abrasion on the rocky shores. The main cable had been made in 2-mile lengths that were later spliced.

Technical consultants for the Atlantic Cable Company included William Thomson (later Lord Kelvin) of Glasgow, Michael Faraday, and Samuel Morse. The board of directors was made up of the three organizers, Brett, Bright, and Field, a number of British bankers and businessmen, and later Lord Kelvin. Dr. Wildman Whitehouse, a doctor of medicine, whose interests had turned to electrical pursuits, was engaged by the company to take charge of testing and other technical matters, and unfortunately he was the superior of the very able consultants who had far more knowledge of telegraph problems.

THE CABLE FLEET

The United States Navy assigned the steam frigate *Niagara* of 5200 tons to the cable-laying project, accompanied by the side-wheeler *Susquehanna*, and the British navy furnished the *Agamemnon* of 3200 tons, accompanied by the paddle-wheel steamer *Leopard*. Both the *Niagara* and the *Agamemnon* had been specially equipped with tanks in their holds for storing the cable, and with cable-laying machinery on their decks, including drums, guide wheels, brakes, dynamometers, and steam engines.

On April 21, 1857, the *Niagara* sailed from New York with Samuel Morse on board. Morse no longer had any financial interest in the Atlantic cable but had an intense personal interest. After his arrival in England, Morse went ashore to renew old friendships.

LOADING AND TESTING THE CABLE

The manufacture of the cable was completed in June, but loading aboard the two ships was not finished until late July. The *Agamemnon*, because of its lighter draft, loaded its cable at the Glass-Elliot works on the Thames, and the *Niagara* took its cargo aboard at the Newall works at Birkenhead, where there was deeper water.

Before going on to Valentia in Ireland, where cable-laying was to begin, the ships met in the quiet water of the Cove of Cork near Queenstown in order to make a final test of the cable. Considering the fact that the cable had been

Figure 9.1 Cable-Laying Machinery on the *Niagara* *(Courtesy Burndy Library)*

made in 2-mile lengths and spliced, such a test was very necessary. It was this matter of splicing which probably accounted for much of the delay in loading. When the ends of the cable aboard the two ships were joined and a battery and galvanometer connected, it was found that the cable was continuous and the insulation was good. The sluggish behavior of a long cable, due to its capacity, was already known and was confirmed by these tests, in which three-quarters of a second was required for a signal to traverse the 2500 miles of cable.

From the Cove of Cork the ships sailed toward Valentia on the southwest coast of Ireland, where they arrived at three o'clock on the afternoon of August 4, 1857. By August 6 the land portion of the cable had been placed and secured.

LAYING THE CABLE

It was agreed that the *Niagara* would begin laying its cable, and in mid-ocean a splice would be made with the cable on the *Agamemnon*, which would then continue the laying to the Newfoundland terminus. The ships moved out on the morning of August 6, but after only 5 miles of cable had been laid it caught in the machinery on deck and snapped. The broken cable end was retrieved in comparatively shallow water and spliced. After a delay of less than a day the convoy again moved westward toward Newfoundland.

During the laying, communication was maintained with the station at Valentia in order to keep a constant check on the condition of the cable. At one point contact was lost for a period of 2½ hours but then resumed. After moving westward for two days, the ocean depth had reached 2000 fathoms and the strain on the cable had increased proportionately, as this great length of cable dangled into the depths of the ocean. It was necessary to keep the cable taut as it passed over the sheave at the stern of the ship, in order to prevent the weight of the cable alone from pulling it out so rapidly as to deposit it in coils on the ocean floor.

Early in the morning of August 11, 1857, the engineer in charge of the drums that paid out the cable found that it was moving out too fast and applied the brake, which grabbed so quickly that the cable broke. At that time 335 miles of cable had been laid, and there was nothing to be done but to return to England. Both ships put ashore at Plymouth and unloaded their cable. At this time the technique of grappling the cable on the ocean floor had not yet been perfected, so that in this instance the cable already laid was abandoned.

PROJECT POSTPONED UNTIL THE FOLLOWING YEAR

The project was given up for the year 1857, but plans were immediately made to continue the project the following year. First of all, 700 miles of new cable were ordered, and substantial improvements were made in the deck machinery for paying out the cable. At this time Thomson developed his marine galvanometer that was to play such an important role in receiving cable messages. With this device it was possible to receive twenty words a minute, or ten times the number previously received. Thomson's galvanometer was very sensitive and consisted of a delicately suspended, short, steel magnet, placed inside a coil of fine, insulated, copper wire. Just above the movable magnet, a small mirror was mounted which reflected a beam of light on a scale placed some distance in front of it. Behind the scale was a lamp the light from which passed through a small opening in the scale to the mirror on the instrument. The reflected

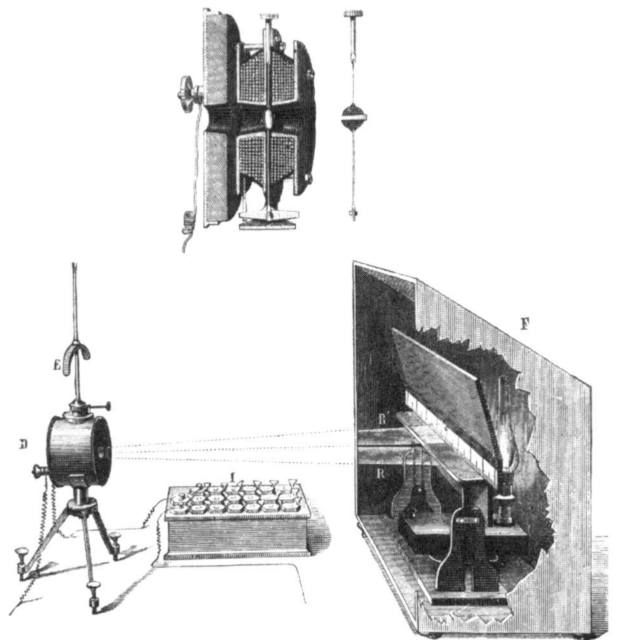

Figure 9.2 Thomson's Marine Galvanometer *(Courtesy Burndy Library)*

spot of light from the lamp appeared on the scale where the deflection could be read.

It was then decided that on the next attempt at laying the cable, the ships would meet in mid-ocean and proceed in opposite directions after making a splice. The winter months of 1857–1858 were devoted to fitting the ships with the new deck machinery, improving the cable tanks in the holds, instructing the crews, and in general making use of experience already gained.

SECOND ATTEMPT

In the spring of 1858, the ships arrived at Plymouth to take aboard the cable deposited there the previous year and in addition the 700 miles of new cable. Loading was completed about the middle of May, and on May 29, 1858, the ships sailed out of Plymouth harbor to the Bay of Biscay, where various experiments and tests were conducted in cable-laying, buoying, and splicing. On June 3, 1858, the ships returned to Plymouth harbor, and a week later sailed out for their mid-ocean rendezvous, where cable-laying was to begin. Besides the *Niagara* and the *Agamemnon*, the supporting vessels, the *Valorous* and the *Gorgon* of the British navy, made up the cable-laying fleet.

By the time the vessels were out on the high seas the greatest storm in the North Atlantic, in the memory of the experienced seamen aboard the ships, broke upon the four vessels. The high seas were particularly hazardous for the *Agamemnon* because she carried a deck load of cable for which there was no room in the hold. After several days the storm abated, and the ships proceeded to their rendezvous without serious damage. When the ships met in mid-ocean on June 25 the weather was calm and all was in readiness.

CABLE IS SPLICED IN MID-OCEAN

On the following day the cables on the two ships were spliced and the joint lowered into the sea. The ships now moved apart; the *Agamemnon* toward Ireland, and the *Niagara* toward Newfoundland. After only 3 miles of cable had been laid, the cable on the *Niagara* caught in the deck machinery and snapped. The ships returned to their starting point, spliced the cable once again, and once more set out toward their respective destinations. On June 27, when the ships were 80 miles apart, the circuit was interrupted by what was found to be two breaks in the insulation. Once more the ships returned to the starting point, and again the cable was spliced. All went well until June 29 when the cable broke about 20 feet from the stern of the *Agamemnon*. About 300 miles of cable was lost, but there still remained a sufficient length to complete the line between the two terminals. The cable fleet turned back and arrived at Queenstown early in July, where the vessels were refueled and loaded with a new stock of provisions. On July 17 the fleet sailed out once again toward the mid-Atlantic. Splicing of the cables was completed on July 29, after which the ships set out for Newfoundland and Ireland. This time only minor difficulties were encountered, and the cable-laying was completed on August 5 by both ships. When the news of the successful cable-landing was telegraphed to New York and from New York to all parts of the nation, celebrations were held everywhere.

INSULATION BREAKS DOWN

This cable continued to operate only until September 1, during which time less than four hundred messages had been transmitted. Various explanations were offered for the failure of the cable. Undoubtedly the insulation had broken down, and some held that the fault lay with the manufacturers, some believed the insulation was injured by exposure to sun and weather while it was stored at Plymouth, and others attributed the damage to the fact that Dr. Whitehouse, in testing, had applied high voltage from an induction coil to the cable.

The clouds of civil war were already hovering over the United States. Cyrus Field's personal fortune was wiped out in a fire in 1859 and a business depression in 1860. In the interim the Atlantic cable management had learned much from the failures, as had also the cable manufacturers.

THE SECOND CABLE

No further attempts were made to lay another cable until the end of the Civil War in the United States. When work was resumed in 1865, a single vessel was employed, the British-owned *Great Eastern*, the largest vessel in the world. The new cable was much improved over those used in the 1857 and 1858 attempts. Among other modifications the conductor was increased in size to seven strands of No. 18 copper wire, with nearly three times the cross section of the earlier cable conductors. Four layers of gutta-percha were used instead of three, and the outer armor of stranded iron wires, of which there were ten, was covered with tarred hemp.

The *Great Eastern* was equipped with three iron-plate tanks for storing the cable and with much-improved paying-out machinery on deck. Below deck was an electrician's room in the charge of Mr. de Sauty, which was equipped with batteries, Thomson's marine galvanometer, clock, log-books, and such other paraphernalia as would be required. Among those on board to render technical assistance were William Thomson, C. F. Varley, and Cyrus Field.

MOST OF THE CABLE IS LAID SUCCESSFULLY BEFORE IT BREAKS

Early in July 1865 the cable was loaded aboard the *Great Eastern*, which then proceeded to the new terminus about 6 miles distant from the former landing place at Valentia. About 30 miles of shore cable had already been placed by the *Caroline*, and attached to a buoy in the harbor. When the *Great Eastern* arrived, the splice was made to the shore cable, and on July 23 she and two accompanying warships of the British navy steamed out of the harbor toward their destination at St. John's, Newfoundland. There were only minor difficulties until a point some 600 miles east of Newfoundland had been reached. At that time a fault developed in the cable after it had passed over the stern of the ship, and an attempt was made to pull it back. As it was being pulled back the cable broke and efforts to retrieve it were unavailing. Although the cable had been grappled three times, in each case either the grappling cable or some of its attachments had failed. Further efforts to retrieve the cable were abandoned and the ships returned to England.

There was of course great disappointment in the minds of all who were connected with the project as well as in the general public. Because of the many failures it became increasingly difficult to raise capital for the cable enterprise. A new company called the Anglo-American Telegraph Company was formed at this time, but it is not entirely clear whether this company was a reorganization of the Atlantic Telegraph Company or merely a holding company to arrange the necessary finances. In any case a new cable was ordered and made ready. This cable again differed somewhat from the earlier ones in that it was stronger, and the armor was made of galvanized wire.

THE THIRD CABLE

The new cable was aboard the *Great Eastern* by the end of June 1866. Two additional ships, the *Albany* and the *Medway* were chartered by the company to accompany the *Great Eastern*. The *Medway* carried what was left of the 1865 cable in addition to 90 miles of heavier shore cable. It was about July 12, 1866, after the shore cable had been laid, that the convoy began its westward journey. Everything worked smoothly until the sixth day out when the cable became badly fouled in the after tank. The ship was stopped quickly, and after several hours the snarl was disentangled. No further mishaps occurred, and the cable was safely landed in a small inlet in Trinity Bay, where there was a tiny village called Heart's Content. The landing of the cable occurred on July 28, 1866, but the work of the expedition was not yet finished. After a brief rest the ships again steamed out into the Atlantic to recover the end of the 1865 cable, a task which was accomplished on August 31. A splice was made, and the second cable was brought ashore on September 7. There were now two cables in good operating condition.

Commercial service on the cables was begun soon after the first cable was brought ashore, using the Thomson marine galvanometer. Reading the messages received by this instrument was necessarily slow, because it depended entirely on observing the deflections of the spot of light on the scale, to the right and to the left. At first the rate of transmission was only eight words per minute, but this rate soon increased to sixteen or seventeen. There was great need for a recording device, but the difficulty lay in the very feeble currents that could be transmitted over the cable. In order to trace a record of the message on paper a frictionless pen was needed.

THE SIPHON RECORDER

In 1867 Thomson, then Sir William Thomson, obtained a patent on what was called the siphon recorder, which was a very different instrument from his mirror galvanometer. Instead of a stationary coil, a very light movable coil was suspended by means of metallic torsion wires between the poles of a strong permanent magnet. Connected to the coil was a small glass siphon tube, one end of which dipped into a bottle of ink and the other end, drawn out to a fine capillary tube, was held a short distance above a brass plate, over which passed the paper strip on which the message was recorded.

The ink bottle was insulated and was kept charged by an electrostatic machine. As the coil moved to and fro, under the influence of the currents from the cable, carrying with it the siphon tube, tiny drops of ink were deposited on the paper in the form of a wavy line, pulled out by electrostatic attraction. The record was very similar to that in Morse's original recording telegraph. There was nothing new in the moving coil principle since it had been applied earlier by Weber and d'Arsonval. The siphon recorder with improvements and modifications became the standard receiving device for cable messages, and was used for many years.

At first the charge for a message between New York and London was five dollars per word, but as the speed of transmission increased and experience was gained in the operation of the cable, the cost was reduced until at the present time such a message costs in the neighborhood of twelve to twenty-three cents per word, depending on the kind of service.

10

The Telephone

BOURSEUL AND REIS

In the August 26, 1854, issue of *L'Illustration* of Paris there appeared an article, written by Charles Bourseul, on the electrical transmission of speech. Bourseul did not claim that he had built an apparatus such as he described. The device that he envisioned had a transmitter consisting of a stretched membrane, to which was attached a metallic contact at its center. When the membrane was caused to vibrate by a sound, its contact would open and close a circuit through a fixed contact in close proximity to the first. The circuit was completed through a battery and a receiver. The receiver consisted of an electromagnet whose poles were very close to a thin iron diaphragm.

Johann Philipp Reis, professor of physics at Friedrichsdorf, Germany, delivered a lecture in 1861 before the Physical Society of Frankfurt in which he described an instrument he had built, which he called the telephone, and which he said would reproduce sounds. Reis's telephone was essentially the same type of instrument described by Bourseul except that the receiver was a knitting needle, wound with fine wire, and attached to a light wood box which acted as a resonator. Reis himself said that he had been unable to reproduce human speech sufficiently distinctly to be intelligible. Others claimed that the instrument would transmit speech, and considered Reis the inventor of the telephone.

Reis and others who worked on this problem apparently did not realize that in order to reproduce speech it was necessary to do more than open and close a circuit very rapidly, but rather that the electric current had to be modulated so as to correspond as closely as possible with the

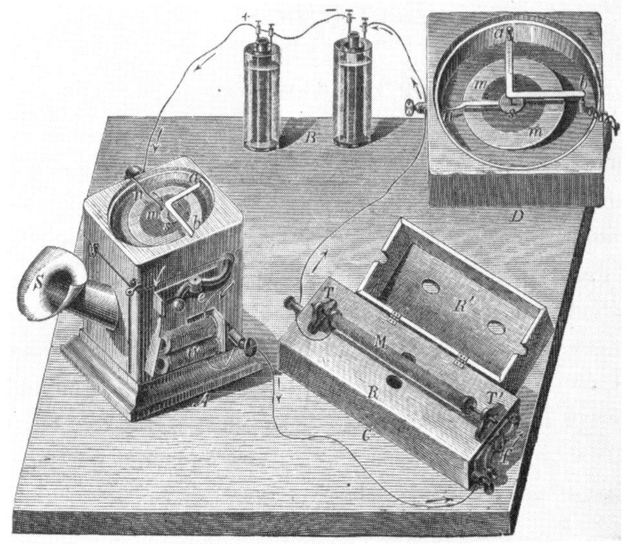

Figure 10.1 Reis's Telephone *(Courtesy Burndy Library)*

complex wave form as well as the volume of the impressed sounds. Helmholtz had shown that the quality of a sound was due to a combination of tones and not to a single pure tone. The Reis transmitter was incapable of transmitting the complex combination of tones that give the human voice or a musical instrument its quality. Furthermore, the Reis instrument could not reproduce the variations in loudness. At best such a transmitter might reproduce a pure tone, but more often the effect on the receiver was merely a rattle.

ALEXANDER GRAHAM BELL

It is generally acknowledged that Alexander Graham Bell (1847-1922) was the inventor of the telephone, but there were numerous contributions by others. Bell was born in Edinburgh, Scotland, on March 3, 1847, to Alexander Melville Bell and his wife Eliza Grace (Symonds). Besides Alexander there were two brothers, Melville James, who was older, and Edward Charles, who was younger than Alexander.

The father had achieved recognition as a teacher of corrective speech and elocution, and also as a lecturer at the University of Edinburgh and at New College. He and his brother published a book in 1860, called *The Standard Elocutionist*, which eventually ran through several hundred editions.

Alexander and his two brothers received their early education from their mother, including lessons in music. At the age of ten Alexander entered a private school in Edinburgh, where he remained for a year, after which he attended the Royal High School, from which he graduated at the age of fourteen. After graduation Alexander spent a year with his grandfather in London, who taught him something of the gracious living of a gentleman and encouraged him in the study of speech and acoustics. Alexander, at this time, had ambitions to become a concert pianist, after having studied the piano in Edinburgh under the tutelage of Signor Auguste Benoit Bertini, but his grandfather discouraged this idea.

When he returned to Edinburgh, Alexander began a more serious study of his father's speech methods and particularly of the applications of *Visible Speech*, which was a method of producing the various speech sounds by means of ten basic symbols involving the positions of the tongue and lips. Combinations of these symbols could produce any speech sound in any language.

At the age of sixteen Alexander became a pupil-teacher at

a school in Elgin. The following summer he studied Greek and Latin at the University of Edinburgh. When he was seventeen he returned to Elgin, where he taught elocution and music at Weston House Academy.

Grandfather Alexander Bell died in London in 1864 or 1865, after which event Bell's father moved to London, leaving his eldest son Melville James in charge of his pupils in Edinburgh. While Alexander was still at Elgin, he wrote a long letter to his father describing some experiments he had conducted with tuning forks, electromagnets, and resonators, in complete ignorance of the experiments of Helmholtz.

The elder Bell became well known in London, and his speech methods were in demand. He was made professor of elocution at the University of London, and altogether had achieved considerable success. In 1868 he was invited to lecture at Lowell Institute in Boston, which invitation he accepted. Alexander, his son, now twenty-one years of age, was left in charge of the London classes, lectures, and lessons.

THE BELL FAMILY MOVES TO CANADA

In 1870 the older brother, Melville James, died of tuberculosis, a disease which earlier had taken the younger brother. Alexander was given a medical examination that disclosed that he too was in some danger. The father, now thoroughly alarmed, decided to abandon his successful London career and to move to Canada. They settled near Brantford, Ontario, where Alexander soon recovered his health, and resumed his experiments with tuning forks and electromagnets.

CLASSES IN BOSTON

In 1871 Alexander was asked to come to Boston to work with the pupils at the School for the Deaf, using his father's method of Visible Speech. After a time he returned to Brantford, but in 1872 he went once more to Boston where he opened a School for Vocal Physiology. Among his pupils was Mabel Hubbard, the daughter of Gardiner Greene Hubbard, a prominent Boston attorney. Mabel had become deaf at the age of four and one-half years after an attack of scarlet fever, and had been sent to Germany to learn lip-reading.

THE HARMONIC TELEGRAPH

From his tuning-fork experiments Bell had conceived the idea of applying the principles he had developed to a device he called his harmonic telegraph. In general the principle of the harmonic telegraph lay in the use of various frequencies, which could be sent over a telegraph line simultaneously and received on separate instruments, each of which was tuned to receive only a single frequency. During the winter of 1872–1873 he carried on a series of experiments with such an apparatus, but he became ill and returned to Brantford in May to recover his health.

BOSTON UNIVERSITY AND GEORGE SANDERS

He did not return to Boston until October 1873 at which time he was appointed professor of vocal physiology in the School of Oratory of Boston University, but despite the new activity he continued his private lessons. At that time also, he became the tutor of a small boy named George Sanders, who lived with his grandmother in a spacious house in Salem. The boy, whose father was Thomas Sanders, had been born deaf. Thomas Sanders was a well-to-do leather merchant in Boston, and was to become Bell's most important supporter, together with Gardiner Greene

Hubbard. Bell was asked to live at the Sanders home in Salem, where he was given ample room for his experimental equipment. He commuted to Boston by train each day and devoted his evenings to teaching the Sanders boy, and after the boy had gone to bed he spent the remaining hours working on his harmonic telegraph. A fine friendship developed between Bell, the Sanders, and the Sanders boy.

THOMAS A. WATSON

Bell was still a British subject at this time and believed that on this account he would be unable to apply for a United States patent. He felt that his harmonic telegraph had reached the stage of development at which it was patentable, but his efforts to secure a British patent at this time were unsuccessful.

His models of the harmonic telegraph were rather crude, and he felt that it was necessary to construct a more workmanlike device. To this end he engaged the services of the Charles Williams shop at 109 Court Street, in Boston. It was here that he first met Thomas A. Watson, a Williams employee, who was to become his assistant and lifetime friend. Watson served Bell in much the same way that Alfred Vail had served Morse. The year was 1874. Bell was twenty-seven years of age and Watson was twenty.

THE PHONAUTOGRAPH AND THE REIS TELEPHONE

Although Bell's primary interest at this time was in the harmonic telegraph, his mind was still occupied with improving his methods of teaching the deaf to speak. He performed experiments with a form of phonautograph, a recent development at the Massachusetts Institute of Technology. This instrument traced the form of a sound wave on a glass plate coated with lampblack. While conducting these experiments at the Institute, he saw there for the first time a reproduction of the Reis telephone. Whether or not he was influenced in his thinking by this model is not known.

From his experiments with the phonautograph, the harmonic telegraph, and even some experiments with human ears obtained from the Harvard Medical School, he gradually evolved the idea of transmitting speech electrically. He knew already that to transmit sounds accurately it would be necessary to produce an undulating current which would follow exactly the complicated waveform of the sound, and that a make-and-break current would not do. These ideas were further developed during a visit to the family home in Brantford in the summer of 1874.

Upon his return to Boston, Bell described his ideas concerning the telephone to Sanders in some detail. Sanders shared Bell's enthusiasm, but Hubbard was not at all excited and wanted Bell to get on with his harmonic telegraph, which he had helped to finance. Bell was now determined to pursue his experiments with renewed vigor, and to that end took over the top floor of the building in which the Williams shop was located.

Bell and Watson worked feverishly during the evenings and at such other times as they might be free from other duties. In 1875 Bell received a United States patent on his harmonic telegraph, a device on which Elisha Gray of Chicago had also been working. Gray was the chief electrician of the Western Electric Manufacturing Company, which at that time produced telegraph equipment and related electrical apparatus for the Western Union Telegraph Company. Gray and Bell were both working on very similar ideas, but Gray had a great advantage over Bell because of his connections with the Western Electric Manufacturing Company, the largest producer of electrical apparatus in the United States.

MEETING WITH JOSEPH HENRY

While Bell was in Washington to obtain his patent on the harmonic telegraph, he met Joseph Henry, the secretary of the Smithsonian Institution who had been the mentor of so many experimenters in electricity, including Morse. Bell mentioned his ideas for the telephone to Henry, but explained that his knowledge of electricity was too meager for the problems involved, whereupon Henry advised him to get the knowledge he needed.

The Western Union Telegraph Company showed an interest in the harmonic telegraph and offered the use of its facilities in New York for further development, but Bell declined the offer. Neither Bell's nor Elisha Gray's harmonic telegraph was developed commercially.

AGREEMENT WITH SANDERS AND HUBBARD

Early in 1875 Bell entered into an agreement with Sanders and Hubbard under which they were to contribute one-half each of the expenses Bell would incur in the development of his telegraph apparatus. No mention was made of any compensation for Bell's time and his living expenses. Bell believed that this agreement covered his work on the telephone as well as the harmonic telegraph, although it appeared later that the other two men who were parties to the agreement thought it covered only the telegraph.

Bell's income from teaching dwindled rapidly as he gave more and more time to his experiments, with the result that he was reduced to great poverty. He had hoped for the sale of his harmonic telegraph to the Western Union Telegraph Company, but in this, as in other things, he was disappointed. His spirits had fallen to a very low ebb, but with Watson's encouragement they worked long hours to improve the harmonic telegraph and a new device, the autograph-telegraph, as Bell called it, or the telautograph, as it was known later.

BELL'S GREAT DISCOVERY

After much fruitless work in April and May 1875, there occurred on June 2, 1875, one of those fortunate accidents that so often have resulted in great discoveries. Bell and Watson were at work on the harmonic telegraph in the attic of the Williams shop with Watson plucking the steel reeds on the transmitter, and Bell tuning the corresponding reeds on the receiver, when Bell became suddenly excited over what he had heard. He ran to Watson to find out what Watson had done. The contacts on one of the reeds of the transmitter had been welded together by the arc so that they failed to open, and when Watson plucked the reed an induced current was produced in the circuit, because the electromagnet under the reed was energized, and an undulating current was produced by electromagnetic induction. This induced current caused the receiving reed to give off a sound unlike that produced when the circuit was made or broken through the contacts. Bell recognized at once that he had discovered the means of producing the undulating current needed to transmit speech.

Bell made a sketch for Watson showing the details of a new apparatus he wanted Watson to build, and by the next day his first telephone was ready. It consisted of two identical instruments, each of which had a membrane of goldbeaters' skin stretched over an opening in a box on the back of which rested a button attached to a hinged iron or steel arm. This arm formed part of a magnetic circuit with a small gap between it and one pole of an electromagnet. The electrical circuit for the two instruments was completed through a battery to energize the electromagnets.

After the Williams shop had closed for the day, the first tests were carried out between the attic and the first floor of the shop over a pair of wires that Watson had strung. Bell had been unable to hear Watson, but almost immediately Watson came running up the stairs exclaiming, "I could hear your voice; I could almost make out what you said." Throughout the month of June, Watson made one instrument after another according to the sketches which Bell prepared, as he sought to make an instrument that would transmit the human voice clearly and distinctly.

Bell felt elated by his meager success, but Hubbard was unimpressed and wanted Bell to perfect his harmonic telegraph and his autograph-telegraph, which in his opinion, were far more likely to produce tangible results and financial returns. Sanders, on the other hand, believed in Bell's ability to make something of real value out of the present flood of confused ideas and projects.

DESPAIR

The summer of 1875 was a wretched one for Bell because in addition to his lack of funds he was in love with Mabel Hubbard, his former pupil. Hubbard was rather put out with him, and his parents disapproved of his neglect of his speech classes. With these burdens pressing in upon him he again fell ill, and once more sought to recuperate at his parents' home in Brantford. As his health returned, his chief concern was the condition of his finances, but he felt that he could not or would not ask Hubbard or Sanders for more assistance. His thoughts turned to some of his father's wealthier neighbors. He called upon George Brown who became interested in Bell's telephone, with the result that he and his brother Gordon Brown agreed to lend Bell the money he needed on the condition that they would be entitled to any foreign patents they might obtain, and that Bell was not to file for a United States patent until such foreign patents, especially the British patent, had been filed, and that Bell would prepare the necessary patent specifications as soon as possible.

Bell returned to Boston and the patent specifications were soon ready, but the written agreement and the money from the Browns did not arrive. Things were getting so desperate that Bell was compelled to return to his teaching in order to live, but he still stood by his agreement not to file for a United States patent until the British patent papers had been filed. The year 1875 slipped away, and there was still no word from the Browns. Hubbard and Sanders urged Bell to file his patent application regardless of his verbal agreement with the Browns. On November 25, 1875, Bell and Mabel Hubbard became engaged, and as the prospective son-in-law of Hubbard his relations with that gentleman improved.

Bell arranged to meet the Browns at Toronto after Christmas, and on December 29 an agreement was concluded under which each of the Browns was to pay Bell twenty-five dollars a month for six months for a half interest in any foreign patents they might obtain.

Bell was working on improving the language of his specifications. One very important modification later saved his patent, namely a clause having to do with a variable resistance, presumably in a transmitter.

George Brown finally sailed for England on January 25, 1876, and Bell as well as Hubbard were in New York to see him off. They waited two weeks, during which time there was no word that the British patent application had been filed. Hubbard's patience had run out, and, without the consent of Bell, he filed for a United States patent on February 14, 1876. It was none too soon, for only a few hours later Elisha Gray filed a caveat with the Patent Office on an almost identical invention.

NEW QUARTERS

At this time Bell moved his equipment from the top floor of the Williams shop building to his living quarters at 5 Exeter Place where he and Watson now carried on their experiments. By this time Bell began to experiment with the electrolytic transmitter, in which a wire attached to a diaphragm dipped into acidulated water contained in a metal cup. On the evening of March 10, 1876, they tried out this transmitter for the first time with excellent results. The transmitter was in one room and the receiver in another, when Bell suddenly shouted into the transmitter, "Watson, come here I want you." Bell had spilled acid on his clothing and wanted Watson to help him. When Bell discovered that Watson had heard his words distinctly on the receiver the spilled acid was forgotten. They changed places back and forth, talking and listening, and both were filled with enthusiasm.

TELEPHONE PATENT GRANTED

Bell's telephone patent was granted on March 7, 1876, and fortunately for Bell he had written his specifications in broad enough language to cover not only his original electromagnetic transmitter but also such other types as the electrolytic.

THE TELEPHONE AT THE CENTENNIAL EXPOSITION

The year 1876 was the centennial of the nation's founding, and in commemoration of that event a great World's Fair was held in Philadelphia. Bell had not entered his telephone, but Hubbard, who was in charge of the Massachusetts educational exhibit, persuaded Bell to exhibit his telephone in that section. When the judges had reached the exhibit nearest that in which the telephone was shown, it was late on a hot day, and the judges were about to leave. Dom Pedro, the Emperor of Brazil, who was with the judges saw Bell, whom he had met previously in connection with his work on deaf mutes, and greeted him warmly. The judges followed Dom Pedro, who, when he learned of Bell's exhibit, insisted that the judges inspect it. One after another of the judges, headed by Sir William Thomson, listened while Bell recited from Hamlet into the transmitter on the opposite side of the hall. At first they were astonished and then became enthusiastic, as they alternately listened and spoke into the apparatus.

They requested Bell to have the telephone moved to the Judges' Hall for further tests. The day was Sunday, and on Monday Bell was to be in Boston for the annual examination of his speech classes, so that it was necessary for someone else to look after the demonstrations. With some misgivings on the part of Bell, William Hubbard was assigned to the task. The young Hubbard acquitted himself very well, so that at least from the judges' point of view the telephone was the most important exhibit at the fair.

TESTING THE TELEPHONE

Late in July Bell went to Brantford once again, taking with him two telephone sets. He made arrangements with the Dominion Telegraph Company for the use of a telegraph line between Brantford and Paris, a distance of 8 miles, for a test of his apparatus. At first he used low-resistance receivers, but the voices were obscured by line noises. When he used high-resistance receivers, the voices came through clearly. The battery that supplied the energy in this case was in Toronto, 68 miles distant, so that the circuit was in reality more than 70 miles in length.

On August 4 Bell's father held a reception at his house in Tutelo Heights, outside of Brantford, and arrangements were made for the use of a telegraph line into Brantford, the last leg of which consisted of iron stove-pipe wire tacked on wooden fence posts to bring the circuit into the Bell home. This test was also eminently successful and delighted the assembled guests. Apparently all of these tests were made with the original electromagnetic telephone in which the transmitter and receiver were one and the same.

During that summer Bell offered Watson a one-tenth interest in the telephone if Watson would devote all of his time to further work on its improvement. Watson was reluctant to accept the offer because he, like Bell, was in need of money, and his job at the Williams shop provided a steady income. After several weeks, however, he did accept and moved into Bell's quarters on Exeter Place. Here the work continued and soon resulted in an important improvement, the substitution of a thin iron disk for the membrane used in previous instruments. A second great improvement, made by Watson, was the use of permanent magnets in the receivers in place of the soft iron cores of the electromagnets. With this instrument batteries were unnecessary.

WESTERN UNION REFUSES TO BUY THE TELEPHONE

Despite his successes Bell felt frustrated, because he wanted to get married and he wanted to repay those who had so generously supported him. With these thoughts in mind he offered his telephone to the Western Union Telegraph Company for $100,000 but was refused.

So far Bell's telephone had brought him no financial returns except for such fees as he collected for his lectures and demonstrations before various organizations, of which there were many. The first actual telephone installation was made between the Williams shop in Boston and the Williams home in Somerville, a distance of three miles.

BELL IS MARRIED

Alexander Graham Bell and Mabel Hubbard were married on July 11, 1877, and early in August they embarked on a steamship bound for England, where they remained for fifteen months. In England, Bell gave demonstrations before scientific and other gatherings, and finally before Queen Victoria. While he was still in England the Electric Telephone Company was organized, and began installing telephones in London, using the Bell inventions. As was the case with Morse, Bell was unable to obtain a British patent because of prior publication. Rival telephone companies sprang up, including one promoted by Fred Gower, whom Bell had employed earlier in connection with his American demonstrations.

ORGANIZATION OF TELEPHONE COMPANIES

In the United States Bell's partners had formed a company called the Bell Telephone Association in which Hubbard, Sanders, and Bell each owned a three-tenths interest, and Watson one-tenth. The company appointed agents in various parts of the country whose duty it was to make telephone leases and make telephone installations. At first these were only private lines, sometimes on the same premises and sometimes between office and home.

Thomas Sanders had invested large sums of money in Bell's telephone up to the point that his credit was becoming stretched, and he was in danger of losing his leather business. In the spring of 1878 Sanders's relatives came forward with the funds needed to organize the New England Telephone Company, which was licensed by the Bell Telephone Association to engage in the telephone business, using Bell equipment.

The Bell telephone at this time was still the simple electromagnetic device in which the same instrument served as both transmitter and receiver. In order to call the person at the other end of the line it was necessary to tap the diaphragm of the instrument so as to produce a faint noise at the other end. Watson made a mechanical device to accomplish the same purpose, which was called Watson's thumper. Shortly thereafter, however, he invented the polarized bell, actuated by a small magneto.

The first crude exchange was installed by the Holmes Burglar Alarm Company in Boston in May 1877. Telephones in various financial institutions that had burglar alarms could be connected through the Holmes central office. The first commercial exchange was put into service in New Haven, Connecticut, on January 25, 1878.

INFRINGEMENT BY THE WESTERN UNION TELEGRAPH COMPANY

In December 1877 the Western Union Telegraph Company, which earlier had refused to buy Bell's patent, organized the American Speaking Telephone Company. It owned the patents on the Edison carbon transmitter, Gray's inventions, and the Page induction coil patent. For the Western Union Telegraph Company to engage in the telephone business was a relatively simple matter because it already owned the necessary lines, the manufacturing facilities, and had ample capital.

With the Edison transmitter and the induction coil Western Union had a telephone far superior to that of Bell and soon made great inroads into the telephone business. The Bell company by this time had about three thousand telephones in service.

Hubbard and Sanders brought suit against the Western Union Telegraph Company for patent infringement. With Bell still in England it was necessary to communicate by cable many times to clarify disputed points. Bell became convinced that it was useless to fight the powerful Western Union Telegraph Company and was completely discouraged.

He left England late in October 1878 with his wife and a small daughter, born to them in May 1878, with the intention of going to Brantford for an indefinite stay. Bell was, however, badly needed to testify at the trial in the infringement suit against Western Union. Watson, who had gone to Quebec to meet the ship which docked on November 10, 1878, succeeded in persuading the reluctant Bell to return to Boston, where he entered Massachusetts General Hospital for an operation. While he was there he made several important depositions of great importance to the infringement proceedings.

BELL PATENT UPHELD

The trial lasted about one year, at the end of which time a settlement was agreed upon under which the patent rights relating to the telephone were to be pooled; the Western Union was to receive a one-fifth interest in the Bell company rentals and four-fifths were to be retained by the Bell company. This settlement was a clear victory for the Bell company in that it established beyond doubt the validity of its patents. One of the factors influencing the outcome was the acquisition by the Bell company of the Blake transmitter in October 1878, making the competitive positions of the two companies more nearly equal.

TRANSMITTERS

Although Bell had experimented with the liquid transmitter in 1876 and found that it yielded much better results than his electromagnetic transmitter and receiver, he clung to his original invention tenaciously until the Bell company was compelled by competition to adopt the Blake transmitter. There were many forms of transmitter in the early history of the telephone, including Reis's metallic contacts, the liquid transmitter, the Emile Berliner transmitter of 1877, the Edison carbon transmitter, the Blake carbon and platinum transmitter, the Hunnings granular carbon and platinum transmitter and its later improvements, and the solid back transmitter invented in 1890 by A. C. White, using carbon disks and granules. The Dolbear electrostatic transmitter was never used in actual practice.

THEODORE N. VAIL

While this book is a history of electricity and magnetism, it is also inevitably bound up with persons and corporations. Up to the time of Bell's sojourn in England the moving spirits behind the Bell companies were Bell, Hubbard, Sanders, and Watson. In May 1878, Theodore N. Vail, who had been general superintendent of railway mails for the United States Post Office, was given the position of general superintendent of the Bell company. It was his hard-nosed attitude which was to a large extent responsible for the favorable settlement with Western Union.

Figure 10.2 Alexander Graham Bell *(From Smithsonian Institution)*

EVOLUTION OF THE BELL COMPANIES

The Bell Telephone Association and the New England Telephone Company have already been mentioned. On July 30, 1878, the Bell Telephone Company was incorporated in Massachusetts, with a capital of $450,000, to succeed the original Bell Telephone Association. On March 13, 1879, the National Bell Telephone Company was incorporated with a capital of $850,000, and it controlled the New England Telephone Company and the Bell Telephone Company. Early in 1880 the American Bell Telephone Company was incorporated in Massachusetts with a capital of $10,000,000, and its stock was exchanged for that of the National Bell Telephone Company, on the basis of six shares for one, after which transaction the latter company was dissolved. Finally the American Telephone and Telegraph Company was incorporated in New York on March 3, 1885, to take over the long lines of the Bell companies, and it eventually became the owner of the majority stock interest in the various regional telephone systems. Its first president was Theodore N. Vail.

Colonel William H. Forbes, together with his brother J. Malcom Forbes, H. L. Higginson, and Lee, Higginson and Company had by May 1, 1880, acquired a controlling interest in the American Bell Telephone Company. Colonel Forbes, who was a close friend of Sanders, had purchased a large block of stock through Sanders in 1879, and in that year became president of the National Bell Telephone Company. When the American Bell Telephone Company was organized, Forbes became its president, with W. R. Driver as treasurer and Theodore N. Vail as general manager.

THE DIAL TELEPHONE

The many improvements made in telephone apparatus in later years are beyond the scope of this book, except that mention should be made of the invention of the automatic or dial telephone, by A. B. Strowger in 1889, which was first used in the La Porte, Indiana, telephone exchange in 1892.

BELL LABORATORIES AND WESTERN ELECTRIC COMPANY

Many inventions and discoveries with uses in and outside the telephone industry, such as the transistor and the solar cell, have been made by the Bell Telephone Laboratories. The Western Electric Company, successor to the Western Electric Manufacturing Company of Chicago and the Williams shop of Boston, a wholly owned subsidiary of the American Telephone and Telegraph Company, has produced telephones, switchboards, cables, and the many other kinds of equipment required in the telephone business.

OTHER TELEPHONE SYSTEMS

The telephone has become almost universal. In most countries outside the United States telephone systems have been largely government-owned and have been connected with the post office. In some countries telephone systems have been installed by foreign private corporations such as the International Telephone and Telegraph Company.

There are numerous telephone manufacturing companies in other countries, especially in Sweden, England, France, Holland, and Germany. In the United States there are several manufacturers other than the Western Electric Company, producing telephone equipment used by independent telephone companies. Although the Bell system controls the greater part of the telephone business in the United States, it is by no means a complete monopoly.

11

Electric Lighting

Three forms of electric lighting have been known for a long time. The electric spark, which later became the electric arc, was the first to be observed. Next came the glow in partially exhausted glass globes excited by an electrical discharge, and finally came the glow of a wire heated by the passage of an electric current.

Hawksbee performed a series of beautiful experiments in partially exhausted glass globes in 1705. Some of these were globes that he used as part of his frictional electric machines. Jean Picard had noted similar effects in the evacuated portion of a Torricellian tube in 1675 but did not know that they were due to electricity. Sir Humphry Davy first exhibited the electric arc at a lecture at the Royal Institution in London in 1809. Current for the experiment was supplied by a large battery that had recently been installed. The dazzling light of the carbon arc made a deep impression upon the gathering.

After the invention of the Leyden jar it was discovered that a discharge from a battery of such jars would heat or even fuse small wires. Soon after Volta's invention of his pile, the heating effects of electric currents were observed. Many of the batteries were sufficiently powerful to fuse iron rods. It is not known who first proposed using such heated conductors as sources of light.

ARC LAMPS

As long as the voltaic battery remained the only source of current electricity it was impracticable to use electricity for illumination because of the expense involved. With the development of magnetoelectric machines and later of dynamoelectric machines, interest in electric lighting sprang up almost immediately. The first practical arc lighting installation, as was previously mentioned, was made at South Foreland Lighthouse in England in 1858, using a Nollet magnetogenerator. Beginning in the middle 1870s, arc lamps and arc lamp machines were produced in great numbers and varied forms.

ARC LAMP MECHANISMS

There were literally hundreds of different arc lamp feed mechanisms. The arc lamp required that the electrodes be in contact for starting and then be separated by a sufficient space to maintain a steady arc of maximum brightness. Usually arc lamps were operated in series so that the same current passed through all of the lamps in the circuit. In order to prevent the entire circuit from being interrupted by the outage of a single lamp, a short-circuiting device was an essential part of the lamp.

Some of the control mechanisms were simple, while others were complicated and ingenious. In general the feed mechanisms employed electromagnets, racks and pinions, ratchets, and gears. Sometimes the upper carbon dropped by gravity as it burned, and in some cases both carbons were moved together mechanically as they became shorter. In most instances the power was supplied by electromagnets, but there were some that contained a clockwork that required winding.

Between the years 1841 and 1844 Deleuil and Archereau devised an arc lamp regulating mechanism that they applied to two lamps installed in Paris. Because these lamps were supplied by Bunsen batteries, the expense of operation proved to be prohibitive. In 1846, however, an arc light was used in a presentation of the opera *The Prophet* in Paris, using a lamp mechanism made by Duboscq. Several inventors in England were also engaged in the design of arc lights, notably Wright and Staite. Staite's lamps were used for lighting a large hall, but the great cost of batteries made all of these attempts impracticable.

CARBONS

The carbons used in the early period of arc lighting were cut from pieces of coke. More satisfactory carbons were made later from a carbon paste molded under pressure. The paste consisted of finely divided coke mixed with a carbonaceous binder. After forming under pressure, the rods were heated to drive off the unwanted gaseous materials and to form a hard carbon rod. A later improvement was made by copper-plating most of the length of the carbon rod to increase its conductivity.

MANUFACTURERS

In Europe the machines of Nollet, Siemens, Wilde, Ladd, Gramme, Schuckert, Ferranti, Hefner and Alteneck, Gulcher, Jablochkov, and others were used for arc lighting. In the United States the principal machines were those of Brush, Thomson-Houston, Weston, and Wallace-Farmer. Each of the American firms mentioned manufactured a complete arc lighting system including lamps, generators, regulating devices, switchboards, and other accessories.

STREET LIGHTING

For twenty years the only arc lamp installations in regular service were the British lighthouses previously mentioned. In 1876 Jablochkov, a Russian, invented a very simple form of arc lamp known as the Jablochkov candle. In this lamp two carbon rods were placed parallel to each other and about ⅛ inch apart. The intervening space was filled with a

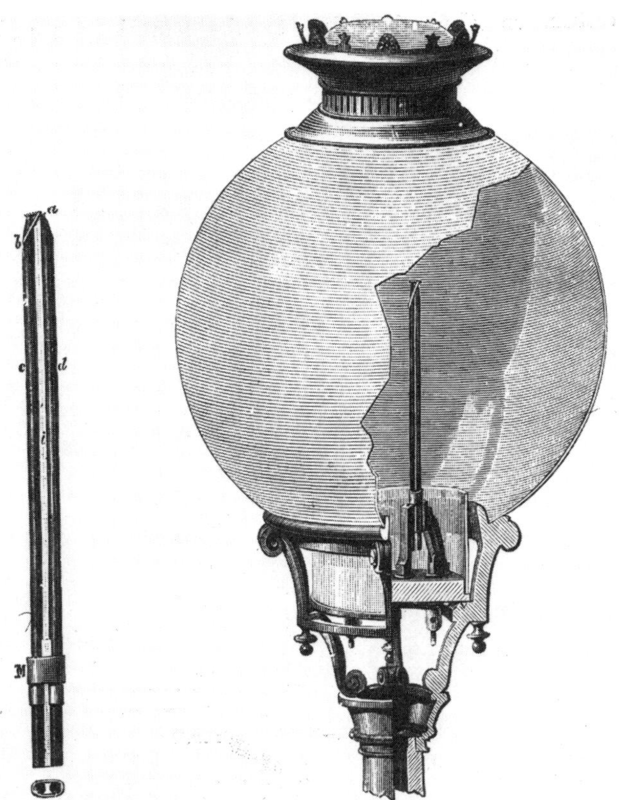

Figure 11.1 The Jablochkov Candle. This improved carbon arc consisted of two carbon rods separated by plaster of Paris. Its great advantage was that no mechanical moving parts were needed to keep a constant arc length. It was used in the first electric street lighting on the Avenue de l'Opera in Paris in 1878. *(Courtesy Burndy Library)*

kaolin cement that acted both as an insulator and a binder. At the top of the candle was placed a powdered carbon mixture or a loose bit of carbon bridging the gap between the rods, so that when the current was turned on an arc was established at each lamp. The Jablochkov system used an alternating-current generator so that the two carbon rods were consumed at the same rate. This was the first commercial use of alternating currents.

The Jablochkov candle gave a soft pleasing light and because of its simplicity gained favor for street lighting purposes. An installation of Jablochkov candles, which operated successfully for three years, was made on the Avenue de l'Opéra in Paris about 1877. Alternating current for these lamps was supplied by Gramme machines. Each lamp contained four candles that were switched on in turn automatically as the preceding candle was consumed. Each candle burned about 1½ hours.

In the United States Charles F. Brush installed arc lamps in the show windows of the Wanamaker store in Philadelphia in 1878. Twenty lamps were employed for this purpose, supplied by five Brush dynamos. Crowds of people gathered to see the new lights, and to them it was a demonstration of a new wizardry of which they had little conception.

Charles F. Brush was a resident of Cleveland, and his lamps, dynamos, and other equipment were manufactured by the Cleveland Telegraph Supply Company. The Wanamaker installation excited the admiration of two youthful professors at the Philadelphia Central High School: Elihu Thomson and Edwin J. Houston. Inspired by the success of the Brush installation, they set about building an arc light machine, with the financial assistance of George S. Garrett.

The Thomson-Houston machine and lamps were successful, and installations of these followed soon thereafter.

An experimental street lighting installation, consisting of twelve Brush arc lamps, was turned on in Cleveland on April 29, 1879. During the same year a company was organized in San Francisco known as the California Electric Light Company, which proposed to go into the business of furnishing arc light service for the streets and buildings of the city. The company's first installation consisted of two Brush dynamos, one with a capacity of six lights and the other sixteen lights. Business grew rapidly so that additions to the plant were needed almost immediately. The first lamps were hung in business places, but within a short time contracts were made for the lighting of streets.

As news of the success of arc lighting spread, there was an immediate demand for electric lights from all parts of the country, so that for several years Brush, Thomson-Houston, Weston, and Wallace-Farmer were unable to keep up with the demand for equipment. By 1884 or 1885 arc lighting for streets was in general use throughout the United States and also in Europe. One interesting application, developed by the Brush company, was the use of tall steel towers, on which were mounted from four to eight arc lights to serve as a sort of artificial moon. Although a great many of these installations were made, they were abandoned within a short time.

Arc lights were well suited to street lighting and for many years were also used successfully for the lighting of large buildings. They were not, however, well adapted to the lighting of homes and smaller buildings, because the light was too intense, it was not of the proper quality, and it flickered considerably. Usually arc lamps were operated on

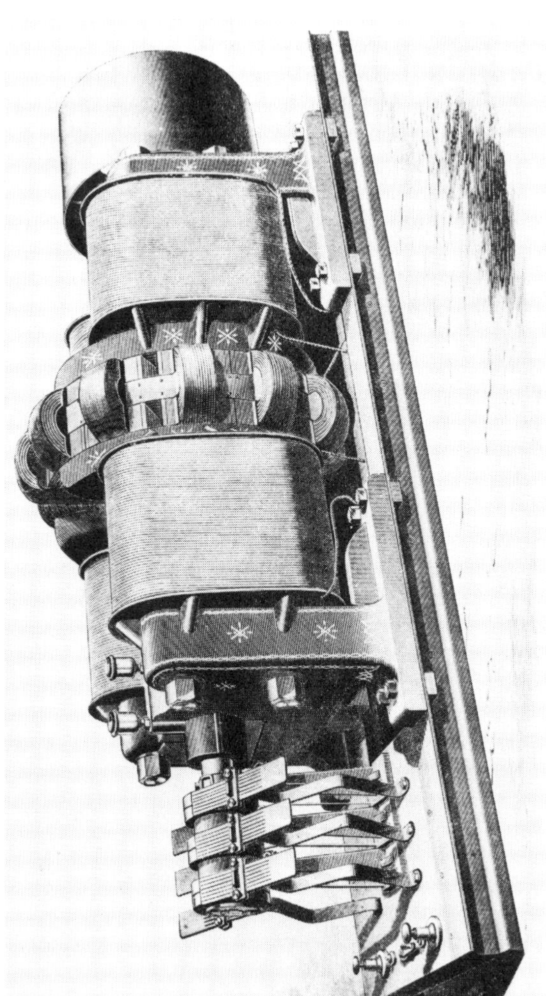

Figure 11.2 An Improved Brush Arc Generator. This dynamo was in use about 1880. *(Courtesy General Electric Company)*

Figure 11.3 Dedication of Lighting Tower. This tower, located in Bridge Square, Minneapolis in 1883, was 257 feet high and carried eight 4000 candle power lamps. As shown in the picture the lamps were lowered for trimming. *(Courtesy Minnesota Historical Society)*

series circuits that required the use of dangerously high voltage.

ENCLOSED ARC LAMPS

There was little change in arc lighting for some years until 1893, when the enclosed arc was introduced by Louis B. Marks. The enclosed arc was used extensively for the interior lighting of stores and larger halls, but it was less efficient than the open arc light. One of its more important advantages was that the carbons lasted ten to twelve times as long as in the open arc.

FLAMING ARCS

In 1898 Bremer of Germany invented the open flaming arc, in which the carbons were impregnated with various metallic salts. It produced intense illumination with colors depending upon the salts used. This lamp had an efficiency of about 35 lumens per watt as compared with 5 to 8 lumens per watt in the enclosed arc and about 15 lumens per watt in the ordinary open carbon arc. A lumen is the quantity of light which falls on a square foot of area of a sphere of 1 foot radius at the center of which is a 1 candlepower light source. Since the area of such a sphere is 12.57 square feet, the theoretical lumens would be 12.57 per candlepower. In practice, because some of the light of any actual light source is blocked off by the lamp itself, the total lumens are sometimes taken to be 10 times the horizontal candlepower.

INCANDESCENT ELECTRIC LIGHTS

Methods of lighting by electricity other than the arc lamp had been tried by various experimenters beginning in the early part of the nineteenth century. The glow emitted by a wire heated by an electric current had been observed soon after the invention of the voltaic battery, and in fact iron and platinum wires had been heated to incandescence or even volatilized by the discharge of Leyden jars.

Sir William Grove, the inventor of the battery bearing his name, devised a lamp in 1840 in which a platinum wire was heated to incandescence by an electric current, while enclosed in a glass tumbler inverted over water.

Apparently the first incandescent lamp for which a patent was granted was that of Frederick de Moleyns of Cheltenham, England. He received a British patent on August 21, 1841, for a lamp in which finely divided carbon was fed into a gap between two coils of platinum wire connected to opposite poles of a battery. The platinum wires and a glass tube containing a supply of powdered charcoal were enclosed in an exhausted glass globe.

J. W. Starr of Cincinnati, Ohio, sought a British patent in 1845 covering two incandescent lamps. In one lamp the light source was a platinum strip enclosed in a glass globe. The length of the platinum strip was variable so as to adapt the lamp for different battery voltages. In the second lamp the substance to be heated to incandescence was a short piece of carbon rod which was held by clamps at each end. This assembly was placed in an expanded chamber at the upper end of a Torricellian tube and therefore operated in a vacuum. Such a lamp could be used for only a short time before the glass became blackened, and like its predecessors it proved impracticable.

Other incandescent lamps were devised by Staite, Nollet, Roberts, Way, Shepard, de Changy, Gardiner, Morris, Farmer, Swan, Lodyguine, Konn, Sawyer and Man, and many others. Moses G. Farmer lighted the living room of his home in Salem, Massachusetts, during the month of July 1859 by means of incandescent electric lights, the current for which was supplied by Bunsen batteries. These lamps, which burned in open air, were merely strips of platinum, tapered at both ends and clamped to terminal pieces.

Sir Joseph Swan's lamp of 1860 had a horseshoe-shaped filament made of a strip of carbonized paper, operating in a vacuum. With the air pumps available at that time the vacuum was rather imperfect, and this fact led to the failure of these lamps. In 1865 the Sprengel mercury pump was invented which made possible high vacuums. In 1878 Swan made lamps with carbon filaments which gave good results. There is considerable justification for the claim that Swan rather than Edison was the inventor of the incandescent lamp, except that Edison produced the first commercially acceptable incandescent lamps and accessories.

EDISON'S INCANDESCENT LAMP

In the United States the same problem was attacked with great energy by Farmer, Sawyer and Man, Maxim, and Edison. Farmer, Sawyer, and Maxim each produced graphite lamps in 1878 all of which operated successfully, but as was the case with all other lamps so far produced they had short lives. The globes soon became black, and they were expensive.

Edison began work on the incandescent lamp in his laboratory in Menlo Park, New Jersey, in 1877. In his first lamps Edison experimented with carbon, then with platinum, with metallic alloys, and then returned to carbon. Several patents were granted to him on the metal filament lamps.

In 1877 and 1878 Edison was for the most part occupied with his work on the phonograph. In July 1878 he accompanied an expedition to Wyoming to view an eclipse of the sun. On his return late in August, he went first to Ansonia, Connecticut, to visit the Wallace Brass works. Wallace and

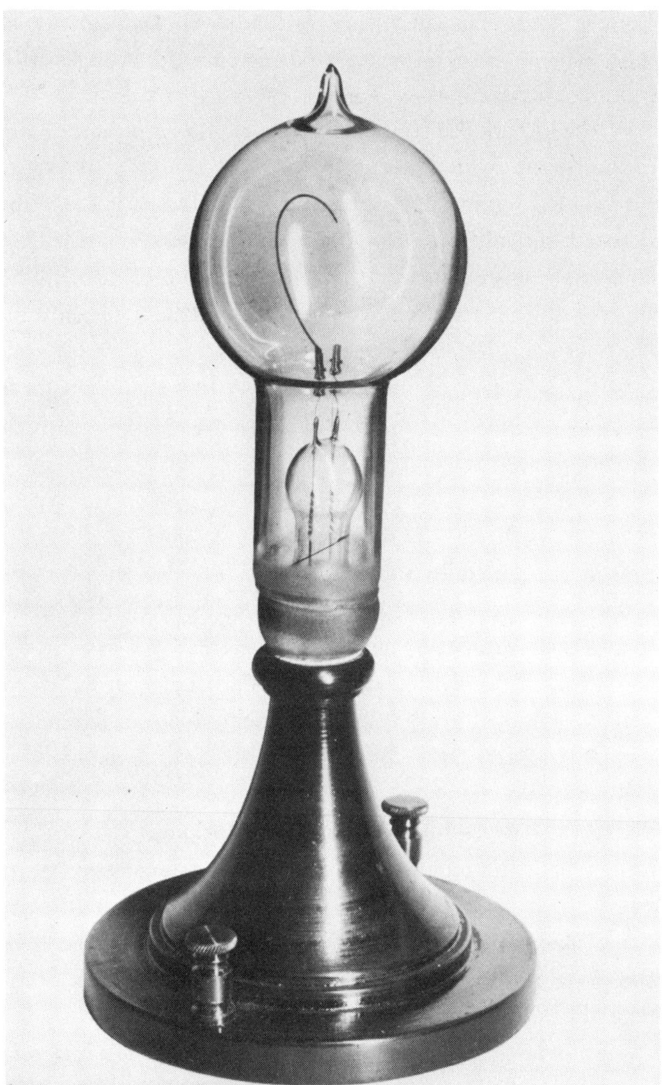

Figure 11.4 Edison's First Commercially Used Lamp *(Courtesy General Electric Company)*

Farmer were manufacturing arc lights and arc light dynamos. When Edison returned to Menlo Park, he brought with him one of the small dynamos.

At Menlo Park Edison now attacked the problem of the incandescent light in earnest. For more than a year he experimented with all sorts of materials, including metals, carbonized thread, paper, wood fiber, bamboo, palm leaves, and whatever else seemed to hold any promise of fulfilling the conditions required for a durable filament. Finally on October 21, 1879, he made a lamp with a carbonized thread filament that burned for forty-five hours before it failed. He applied for a patent on November 4, 1879, which was granted on January 27, 1880.

EDISON ELECTRIC LIGHT COMPANY

Edison was not only a great inventor but also an excellent businessman and an organizer. On October 17, 1878, more than a year before he made his first successful incandescent lamp, The Edison Electric Light Company was organized with a capital of $300,000. The stockholders and directors included some of the most important New York bankers and financiers.

After his successful carbonized thread lamp, Edison continued his experiments and made rather a large number of lamps with carbonized paper filaments. Incidentally, the name filament was first applied by Edison. Before that time the glowing elements in the lamps had usually been designated as burners, to correspond with gas burners.

On Sunday, December 21, 1879, the *New York Herald* printed a full-page description of Edison's lighting system, which by this time included not only his incandescent lamp but also his dynamo and auxiliary equipment. After the publication of the article in the *Herald*, gas company stocks fell rapidly in price and Edison Electric Light Company shares were traded for as much as $5000.

MENLO PARK

On New Year's Eve 1879, an experimental installation of lamps with carbonized paper filaments was made at Menlo Park. Special trains were run from New York City to accommodate the crowds of people who came to see the new lights. Edison had succeeded in "subdividing" the electric light according to the queer expression that had become current during the preceding period. For some unaccountable reason people had the notion that the electric light was the arc light, and it was argued with great vehemence that the light could not be subdivided, that is, produced in smaller units of lower intensity than the familiar arc lamp.

THE SEARCH FOR BETTER FILAMENT MATERIALS

Despite his successes Edison was not satisfied that he had obtained the best possible material for his lamp filaments, and continued his experiments with various vegetable fibers. He dispatched emissaries to Japan, China, Burma, Ceylon, and South America to obtain specimens of bamboo, grasses, palms, and wood. His greatest success was with bamboo, and for nine years his lamps were made with carbonized bamboo filaments.

IMPROVEMENTS IN LAMP SEALS AND IN DYNAMOS

He was also working on various other problems connected with the incandescent lamp such as improving the seal around the lead-in wires, obtaining a better vacuum by expelling the occluded gases in the filament, and making a

Figure 11.5 An Edison Bipolar Dynamo. This generator was developed in 1886. *(Courtesy General Electric Company)*

more uniform filament. In some of these endeavors he was aided by the discoveries made by others.

Edison was engaged also in developing a constant-voltage dynamo, with low internal resistance and greatly improved efficiency. Many of the earlier machines had efficiencies of only 50 to 60 percent, but Edison succeeded in building a machine with more than 90 percent efficiency.

FIRST COMMERCIAL INSTALLATIONS

The first commercial installation of Edison's new lamps and dynamos was made on the steamship *Columbia* which was under construction in the shipyards at Chester, Pennsylvania, for the Oregon Railway and Navigation Company. This installation included four dynamos, three of which operated in parallel, with the fourth used as an exciter. This plant went into operation on May 2, 1880. Early in January 1881 an Edison lighting system was installed in the lithographing plant of Hinds, Ketchum and Company at 229 Pearl Street, New York. About 150 installations in private plants were made during the following two years with more than 30,000 lamps.

Edison's lamps by this time were equipped with the now familiar screw base. As the production of lamps increased in quantity, the quality improved and the cost was greatly reduced. The growth of the lamp business soon made the facilities of the Menlo Park laboratories inadequate. A factory building was purchased in Menlo Park in 1880 for use as a lamp factory, but this also soon became too small, so that in 1881 a group of factory buildings in Harrison, New Jersey, was taken over by Edison for the manufacture of lamps. This plant was operated by the Edison Lamp Company, a separate company that had been organized in 1880.

PEARL STREET, THE FIRST CENTRAL STATION FOR INCANDESCENT LIGHTING

Edison had realized for some time that substantial economies were possible through pooling the generating facilities for a large number of customers in a single plant, or in other words through the construction of a central station. Savings would result from what is now known as diversity among the customers and also from greater efficiency resulting from the larger size of generators and lower labor costs.

Arc lights had already become familiar sights in New York City with three-quarters of a mile of Broadway illuminated by the new lights in December 1880. On the twentieth of that same month the Edison Electric Illuminating Company of New York was incorporated. This company made a contract with the Edison Electric Light Company for a complete installation of a generating station, underground cables, wiring of buildings, lamps, fixtures, meters, and all other auxiliary equipment, for incandescent light service in an area of about one-sixth of a square mile, bounded roughly by Wall Street, Nassau Street, the East River, and Spruce Street.

The plant was constructed at 257 Pearl Street, with boilers at the street level and engines and dynamos on the floor above. Engines and boilers were purchased from the manufacturers of such equipment. Dynamos were built by the Edison Machine Works on Goerck Street in New York City. Sockets, switches, and other equipment were produced by Bergmann and Company on Wooster Street. Bergmann had been associated with Edison at Menlo Park. Underground conduits and fittings were made by the Underground Tube Company at 65 Washington Street.

The Pearl Street plant and distribution system were completed and put into operation on September 4, 1882. The system was an unqualified success notwithstanding many misgivings of Edison and his associates. This was the first incandescent central station in the United States, but there had been many earlier stations supplying arc lamp service.

The success of the Pearl Street station resulted immediately in orders for similar systems from all parts of the country. The load on Pearl Street grew rapidly, so that the

Figure 11.6 The Dynamo Room at the Pearl Street Station. This was Edison's first central station for incandescent electric lighting. It began operation in New York City in 1882. *(Courtesy General Electric Company)*

Figure 11.7 Thomas A. Edison *(Courtesy General Electric Company)*

requests for service soon exceeded the capacity of the station. In the original installation there were six Jumbo dynamos capable of supplying 7200, 16-candlepower lamps.

SCHENECTADY WORKS

In order to fill the flood of orders for all types of equipment the various Edison factories expanded enormously. In 1886 the Edison Machine Works moved to Schenectady, New York, where it later became the nucleus of the great General Electric Company works. Harrison, New Jersey, remained the site of one of the greatest lamp factories.

FOREIGN INCANDESCENT LIGHT INSTALLATIONS

The Paris exposition of 1881 gave great impetus to the adoption of incandescent lighting. At this exposition Edison exhibited a complete lighting system, including generators, conductors, switches, fuses, lamps, sockets, and fixtures. Other exhibitors were Swan and Lane-Fox of England, and Maxim of the United States. In 1882 there was an electrical exposition in the Crystal Palace in London that included many electric lighting exhibits and a complete installation by Edison of his generating equipment, lamps, sockets, and accessories.

At about the same time the Edison Electric Lighting Company of London Ltd. was constructing a plant in a building at Holborn Viaduct to serve nearby customers, among whom was the British Post Office. This plant began operation in March 1882, about six months earlier than the Pearl Street station.

Early Edison lamps had no base, but the lead-in wires were merely bent back over the neck of the lamp and made contact with two spring clips in the socket. In the latter part of the year 1880 the first Edison screw base lamps appeared, followed by improvements in 1881.

Many other manufacturers of incandescent lamps appeared in the 1880s, including Thomson-Houston, Westinghouse, Brush-Swan, Hawkeye, Fort Wayne, Perkins, Loomis, and Schaeffer in the United States, Swan in England, and Siemens and Halske in Germany. Each of these manufacturers used a different lamp base and socket. Gradually, however, there was a shift to the Edison screw base, and in many cases, in which a different socket had been installed originally, adapters were provided so that Edison lamps could be used. In the early 1900s the Edison base became the standard for the United States except for special applications. Many European manufacturers also adopted the Edison base, but some, even to the present day, use a bayonet type of base, similar to that used on automobile lamps.

IMPROVED LAMPS

The bamboo filament lamp was produced in great numbers and was found to be rugged, but its efficiency was only about 1.7 lumens per watt when new, and only half that amount when the bulb had become blackened. With constant research and consequent improvements the efficiency reached 2.5 lumens per watt in 1881. There was also a gradual improvement in manufacturing methods, in lamp construction, in lamp life, and a substantial decrease in cost. In 1890 or thereabouts, there was a significant change when the squirted filament was introduced. This filament, as the name implies, was made by squirting a carbonaceous paste through a die, after which the soft thread was carbonized. Lamps of this type and the improved Gem lamp of 1905 were manufactured until about 1918, but long before that time metal filament lamps had taken over the great bulk of the market.

OTHER TYPES OF LAMPS

Many new types of lamps came into existence beginning at the turn of the century. Aside from the arc lamp improvements previously mentioned, came the low-pressure mercury arc lamp invented by Peter Cooper Hewitt in 1901, and with it the mercury arc rectifier. Also in 1901, H. W. Nernst of Germany invented the Nernst lamp in which the light source was composed of one or more rods of molded oxides of zirconium and yttrium. This material was practically a nonconductor when cold but had an increased conductivity with increase in temperature. It was necessary therefore to provide a heating coil to raise the temperature sufficiently so that current would flow through the glower itself. This lamp was used for only a few years for indoor and outdoor lighting.

The osmium lamp was developed in 1905 by Dr. Auer von Welsbach, followed by the tantalum lamp in 1906 by Dr. Werner von Bolton, both of Germany. The sintered tungsten lamp was patented in Germany by Just and Hanaman of Vienna in 1903, and a United States patent was applied for in 1905, but because of other claims the patent was not granted until 1912.

While tungsten was an excellent material for lamp filaments, the metal was so brittle that it had not been possible to draw it into wire. The filament produced by the Just and Hanaman method was formed by taking finely divided tungsten, mixing it with a binder, squirting the paste through a die, and sintering the particles of tungsten together by passing a current of electricity through the thread that had been formed. Lamps made by this method were far more efficient than carbon lamps, but the filament was weak and easily broken, and the lamps were expensive. Nevertheless the sintered tungsten lamps went into production in Hungary in 1905 and in the United States in 1907.

The problem of making drawn-wire tungsten filaments was finally solved by Dr. William D. Coolidge of the General Electric Company in 1911 after about four years of work. In a relatively short time drawn-wire tungsten filament lamps were produced for the market. In 1913 Dr. Irving Langmuir, also of the General Electric Company, perfected the gas-filled tungsten lamp. This lamp was a great step forward because the filaments in these lamps could be operated at considerably higher temperatures and consequently higher efficiencies, particularly in the larger sizes, where efficiencies of 20 lumens per watt were reached.

TUBE LIGHTING

Geissler tubes, which were mentioned previously, were produced in great numbers around the middle of the nineteenth century. Some of these tubes contained mercury or various of the rare gases in small amounts, which gave off their characteristic color spectra when excited by the discharge of an electrostatic machine or an induction coil. Many of the tubes were also made with glass containing salts that fluoresced in beautiful colors when they were excited.

About the year 1895, D. McFarlan Moore of the United States began experimenting with long glass tubes filled with carbon dioxide gas, which gave off a good quality white light when a current of electricity was sent through them at relatively high voltage. Beginning at about the year 1904, many installations of such tube lighting were made, especially in stores.

Geissler had used neon and other rare gases in his tubes, but such gases were expensive to produce. Georges Claude of France discovered a method of obtaining neon cheaply, and about 1910 began making neon tube lighting, but it was not until 1922 that such lighting was introduced commercially, largely for advertising.

The principle underlying sodium vapor lamps was discovered about 1910, but they were first used for highway lighting in 1933.

High-pressure mercury lamps were developed in 1912, but these lamps, now widely used for highway and airport lighting, were not introduced commercially until 1934.

FLUORESCENT LAMPS

The early research work on fluorescent lamps was done in Germany and in England. It had long been known that certain minerals would fluoresce under the influence of ultraviolet light, but no practical application was made of such phenomena until many years later. The first fluorescent lamps were made in Europe in 1934. They were introduced commercially in the United States in 1938. The fluorescent lamp is essentially a low-pressure mercury arc lamp, producing ultraviolet light, which in turn causes a coating on the inside of the tube to fluoresce with a color depending upon the kind of material used.

LAMP EFFICIENCIES

Efficiencies had increased enormously with the introduction of new types of lamps. Sodium vapor lamps had efficiencies of about 70 lumens per watt. High-pressure mercury lamps yielded from 40 to 55 lumens per watt, and fluorescent lamps produced a lifetime average of 51 lumens per watt. These figures may be compared with the early Edison lamps that yielded less than 2 lumens per watt.

SPECIAL-PURPOSE LAMPS

No radically new types have been introduced since the fluorescent lamp, but there has been a steady improvement in most existing types. Special-purpose lamps are now available in hundreds of varieties, including many miniature bulbs, automobile lamps, sterilamps, sunlamps, photographic lamps, glow lamps, and a host of others.

12

Alternating Currents

The early electric lighting systems, both arc light and incandescent, used direct current with the notable exception of Jablochkov candles, which used alternating current so that the electrodes might burn at the same rate. Some of the first magnetogenerators, such as the Nollet machines, had been built as alternators.

Electrical designers and manufacturers of the early 1880s were not altogether ignorant of alternating currents, but from the viewpoint of that day direct current offered so many advantages that alternating current had not been considered seriously. Among other things there was no alternating-current motor, and alternating-current generators could not be connected in parallel as Edison was doing with his dynamos. Direct current was suited to electroplating, electrolytic processes, and could be used for charging storage batteries. Arc lights performed better and more quietly on direct current, and many electromagnetic devices would not function satisfactorily on alternating current.

Despite all of these disadvantages there was nevertheless one great advantage that was recognized after a time, namely that alternating currents could be readily and efficiently changed from one voltage to another, and since higher voltages were essential for transmission over distances greater than about ½ mile, it was inevitable that alternating currents would be employed.

THE TRANSFORMER

The heart of the alternating-current system was the transformer, and when the transformer had been brought to a satisfactory stage of development the alternating-current system was established. The transformer was not really new in principle because Faraday in his original discovery of electromagnetic induction had used what was essentially a transformer: a ring of iron with several insulated coils of wire wound upon it. Joseph Henry made coils of flat copper ribbon wound upon an iron core with which he was able to produce strong sparks and shocks due to self-induction. These experiments by Henry were published in 1835. During that same year, and following the path of Henry's experiments, Professor C. G. Page of Washington made up similar coils on which connections or taps were made at intervals throughout their length. Page found that when a battery circuit was made or broken through any two of these connections, sparks could be obtained through any other pair of connections. This arrangement was in effect an autotransformer.

INDUCTION COILS

The Reverend N. J. Callan of Maynooth College in Ireland constructed an electromagnet, as he called it, a description of which was published in 1836. It had a horseshoe-shaped iron core upon which he wound two coils of wire, one upon the other. The inner coil was of heavy copper wire 50 feet long, and the outer coil was of fine iron wire 1300 feet long, each turn insulated from its adjacent turns. The two coils were connected and a tap was brought out from the junction between the coarse and fine wire coils.

Callan found that by making or breaking a battery circuit through the inner coil strong shocks could be obtained from the outer coil. In the same year Callan constructed a much larger coil with two entirely separate windings, of which the inner one was copper wire 1/6 inch in diameter, and the outer one also copper wire 1/40 inch in diameter and 10,000 feet long. He provided this coil with an interrupter to produce an intermittent current. This coil produced sparks of considerable length. In 1837 Callan made several coils with separate primary and secondary windings, and he was able to increase the effect by connecting the secondaries of several such coils in series.

In the same year Sturgeon wound a double coil on a hollow wooden spool into which he inserted iron cores. Professor C. H. Bachhoffer, in working with a similar coil, discovered that a core made up of a bundle of iron wires gave much better results than a solid iron core.

Page made several two-winding induction coils in 1838, using iron wire cores and magnetic interrupters. Page asserted that he had used iron wire cores before February 1838, so that there is some uncertainty as to whether he or Bachhoffer deserves the credit for the invention. The interrupter used on these coils was Page's invention. It was actuated by the magnetism in the core and the break was made in mercury cups.

Wagner and Neef made a modification of Page's interrupter in which platinum contacts were substituted for the mercury cups. MacCauley of Dublin may, however, have made a similar interrupter at an earlier date than Wagner and Neef.

In France Masson and Bréguet began to make induction coils of excellent workmanship in 1838. With them they charged Leyden jars and produced luminous discharges in evacuated tubes. Similar experiments had, however, already been performed by Page.

H. D. Ruhmkorff of Paris began making induction coils about 1851. He was familiar with the work of Masson and Bréguet, and others who had already produced powerful coils. One of these built by Page in 1850 gave a spark of 8 inches. Ruhmkorff was an excellent craftsman, and made coils of great refinement and excellence, so that the term Ruhmkorff coil was synonymous with induction coil for many years. In order to overcome the difficulty of internal sparking, Ruhmkorff divided the secondary winding into sections, which were well insulated from each other and from the primary winding.

Fizeau, whose fame is due chiefly to his work with light, discovered in 1853 that the performance of an induction coil was greatly improved by connecting a condenser around the points of the interrupter. Such a condenser served to absorb the self-inductive discharge from the primary and made the break more abrupt. It served also to reduce the burning of the points by minimizing the spark at that contact.

Ruhmkorff adopted the condenser and made many improvements in various details, so that his coils were probably the finest made anywhere. He engaged in their manufacture on a large scale and was therefore often considered as the inventor of the induction coil. One of his largest coils made in 1867 gave sparks of 40 centimeters, or about 16 inches.

Another Parisian, named M. Jean succeeded in obtaining a 30-inch spark by immersing his coil in turpentine. It is possible however that Poggendorff may have originated the idea of oil immersion. Poggendorff improved the condenser by using mica in place of oiled silk or paper. He studied carefully the proportions of coils that give the best results and the size of condenser required for coils of different sizes.

Other important coil makers were Störer of Germany, Apps and Grove of England, and Ritchie of the United States. Ritchie made a coil in 1860 that gave a spark of 53 centimeters. The most famous coil was one built by Apps for Mr. Spottiswoode, known as the Spottiswoode coil, which gave a spark of 42 inches, or nearly 107 centimeters.

Sir William Grove used alternating current from a small magneto machine to excite an induction coil. Up to this time the magnetic circuits of induction coils were open. The cores, except those of Callan, were almost invariably cylindrical in form. In 1856 C. F. Varley obtained a British patent for an induction coil with a closed magnetic circuit. The iron wires that made up the core were considerably longer than the coil. When the coil had been completely wound and insulated, these wires were bent back over the coil from both ends so as to encase it completely, and were then firmly secured by bands.

Jablochkov had applied for a patent in 1877 for a coil to be used in connection with his arc lighting system in Paris. British patents were issued in 1878 to H. Wilde, C. W. Harrison, C. T. Bright, J. B. Fuller, and de Meritens for induction coils, or transformers to be used for purposes such as lighting and telegraphy. Marcel Desprez and Jules Carpentier in France proposed in 1881 to use an induction coil to raise the potential at the generator for transmission to a more distant point and then to use a second induction coil, with connections reversed, to reduce the potential to a value suited to the purpose for which the energy was to be used. This proposal seems to have been the first to state clearly the use of transformers as we now know them.

GAULARD AND GIBBS

Gaulard and Gibbs began work in England in 1882 on the application of induction coils, or transformers, or secondary generators as they called them, to long circuits for lighting. Their original coils had straight cylindrical cores of solid iron. They were used on series circuits with the primary circuits connected in series, following the method used by Jablochkov.

At an electrical exhibition held in London in 1883, Gaulard and Gibbs exhibited two of their coils, connected in series to a Siemens alternator. One of these supplied the energy for a group of incandescent lamps, and the other lighted several Jablochkov candles and incandescent lamps and also operated a small motor, probably a series commutator machine. Later in 1883 Gaulard and Gibbs made an installation along the Metropolitan Railway in London. Current was sent over a series circuit 16 miles in length through four transformers that served incandescent and arc lamps at four of the railway stations.

At an exhibition held in Turin in 1884, they set up a circuit 50 miles in length with four of their secondary generators along the route. Each of these served a group of arc and incandescent lamps. Gaulard and Gibbs were refused a British patent but were granted a United States patent in 1886. Their transformer had before this time been redesigned with a closed laminated magnetic circuit.

In the meantime, during the year 1885, Zipernowsky, Déri, and Bláthy of Budapest exhibited a pair of transformers in London with closed magnetic circuits. These were operated at 1000 volts on the primary side and 100 volts secondary. Other more or less successful transformers with closed magnetic circuits, had been produced by Rankin, Kennedy, and Hopkinson.

William Stanley had been experimenting with transformers, which he called converters, during the years 1884 and 1885. In the latter year, working independently of developments abroad, he set up a circuit at Great Barrington, Massachusetts, which included an alternator, a ½-mile line, and a transformer, from which he operated 150 16-candlepower lamps. Stanley's transformer was of the shell type; that is, the iron core to a large extent encircled the windings. The core was made of sheet iron stampings, suitable for manufacturing processes.

WESTINGHOUSE ALTERNATING-CURRENT SYSTEM

George Westinghouse had been watching the development of the alternating-current system with great interest and in 1885 purchased the Gaulard and Gibbs rights in the United States. He engaged William Stanley to work on the problem and to design a transformer that could be produced in quantities by ordinary manufacturing processes. The Westinghouse Company began the manufacture of alternating-current apparatus within a short time, including transformers, generators, and auxiliary equipment. The first large-scale installation was made in Buffalo in 1886. In the following two years a great many alternating-current installations were made by the Westinghouse Company in many cities throughout the United States.

There was considerable opposition to alternating currents, especially by the Edison Company, based on the fact that high voltages were employed, which it was said were very dangerous, as they probably were at the time. The Thomson-Houston Company hesitated for several years, and then in 1888 brought out a competing alternating-current system. Professor Thomson had experimented with transformers while he was still teaching at Central High School in Philadelphia in 1879 and had delivered several lectures on the subject.

ALTERNATING-CURRENT GENERATORS

Little has been said about alternating-current generators. They developed quite naturally from the direct-current dynamo, but with some modifications. Whereas the direct-current machines at the time were almost entirely bipolar, alternators were built as multipolar machines in order to obtain the frequencies necessary to prevent flicker in incandescent lights at moderate speeds. Prime movers were mostly slow-speed steam engines, and even though generators were belted, their speed was still slow. The speed of direct-current machines was kept low for another reason. In the early days, brush and commutator troubles were among the greatest difficulties experienced by the operators, and these troubles were greatly exaggerated at higher speed. When the carbon or graphite brush was introduced, some of the difficulties disappeared.

FREQUENCIES

Frequencies at first were generally rather high; 133 cycles was common. With the coming of the induction motor, however, it was found that lower frequencies were far more suitable for power. Because of these conflicting requirements the matter of frequencies was for a time in a chaotic state. Gradually the higher frequencies disappeared and most systems settled down to 50 or 60 cycles for lighting and 25 cycles for power, but there were some exceptions.

AC-DC CONVERSION

Most electric utilities started out with Edison or direct-current systems in their business sections, but as the demand for electric service for residential use developed, alternating-current systems were built to serve the more distant customers. These alternating current systems at first were single phase and were later replaced by two- or three-phase installations.

With this combination of direct- and alternating-current

systems, it became desirable to shift the load from one to the other. For this purpose the rotary converter was the simplest answer, but with the knowledge then available rotary converters for the higher frequencies were impracticable because of the high commutator speeds. It was not until about 1905 that 60-cycle rotary converters were introduced. Until that time they were used on 25-cycle or slightly higher frequencies, and motor-generators were used at higher frequencies.

Street railway systems were changing from horse-drawn to electric traction during the late 1880s and the early 1890s under the leadership of Siemens and Hopkinson in Europe, and Edison, Sprague, Van Depoele, Bentley and Knight, and others in the United States. The advantages of alternating current generation, transformation, and transmission, with conversion equipment in substations at various points soon became apparent. Here also, the lower frequencies were necessary because of the limitations of conversion equipment. In many cases the railway companies' equipment was more modern than that of the electric utilities because the railway companies had the advantage of the experience gained by the electric companies.

ALTERNATING-CURRENT MOTORS

The direct-current motor was already well established when the alternating-current system came upon the scene. Series motors sometimes operated on series arc light circuits, and shunt or compound-wound motors were used on the incandescent systems. It was found that some of the series motors performed tolerably well on alternating current, but at reduced capacities.

A form of induction motor was described by F. C. Bailey in 1879. Professor Galileo Ferraris of Turin, Italy, developed an experimental induction motor in 1885 that he described the following year. Apparently the purpose of his invention was not really a motor but rather a meter for alternating currents.

In May 1888 Nikola Tesla announced, in a paper read before a meeting of the American Institute of Electrical Engineers, his invention of the induction motor. George Westinghouse realized immediately the value of the new motor and acquired forthwith not only Tesla's invention but also his services. Putting the new motor into commercial service was not without its difficulties. The early alternating-current systems were single-phase, 133 cycles—too high for a motor. The self-starting single-phase motor was not available at first, and without two- or three-phase circuits, the induction motor was useless.

The coming of the induction motor, therefore, made necessary the overhaul and redesign of much of the existing apparatus and the scrapping of many machines only two or three years old.

NIAGARA FALLS DEVELOPMENT

There were widespread effects from the development of an alternating-current motor, most important of which was the power development at Niagara Falls. The original charter for power at Niagara Falls was granted in 1886. It was planned at that time to construct a tunnel to receive the water discharged from hydraulic turbines located at various points in the city of Niagara Falls and discharge this water at a point below the falls. Industrial sites were to be provided along a canal from which each lessee would take the water for which he contracted and discharge the tail water from his turbines into the tunnel.

Although the generation of electricity had been considered, no serious thought had been given to converting large amounts of waterpower into electrical energy. The project was under discussion for several years, and many solutions were offered, including the transmission of power by compressed air. Fortunately the development of alternating-current apparatus had progressed sufficiently by this time so that its use was considered practical. The solution finally adopted, after bitter debate and with considerable doubt and misgivings by many, was the plan offered by Professor George Forbes, which called for the installation of polyphase alternating-current generators.

Work on the project was begun on October 4, 1890. The contracts for electrical equipment were awarded to the Westinghouse Company, and called for three 5000-horsepower, three-phase, 25-cycle, 2200-volt machines. This contract was let in October 1893, with delivery in about a year. Considering the fact that these were the largest machines built up to that time, and considering also that these were vertical machines in which the entire weight of the rotor was suspended from the top bearing, it was an engineering accomplishment of the first order. The first power was delivered in August 1895, and the first customer was the Pittsburgh Reduction Company, which later became the Aluminum Company of America. This company produced aluminum by the Hall process, which had been discovered in 1886. Operation of the first installation was eminently successful, and within a short time seven additional generators were installed in Power House No. 1.

TRANSMISSION LINES

In the year 1891 a notable achievement was recorded in Germany with the construction of a 110-mile three-phase transmission line between Lauffen and Frankfort. The exact voltage used on this line is somewhat in doubt with reported figures ranging from 8500 to 30,000 volts. The most probable voltage is 16,000. In the United States lines were built from Willamette Falls to Portland, Oregon, about 1892 at 5000 volts, and from San Antonio, California, to San Bernardino, also in 1892, 28 miles, with a 15-mile tap to Pomona. This second line operated at 10,000 volts single-phase.

The success of these early lines encouraged the engineers at Niagara to undertake the transmission of power to Buffalo in 1896. The first line was 22 miles in length and operated at 11,000 volts, three-phase. After Power House No. 1, with a capacity of 50,000 horsepower, had been completed, Power House No. 2, with an identical 50,000 horsepower, was begun, but the electrical contracts were awarded to the General Electric Company.

Following the success of the Niagara Falls development, power companies throughout the country began the installation of two- and three-phase systems to replace the now-outmoded single-phase equipment. Other large-scale hydroelectric developments followed close upon the heels of the Niagara Falls installation.

FREQUENCY AND VOLTAGE STANDARDS

Because the electrical age had come so suddenly there were as yet no generally accepted standards. Not only were there direct- and alternating-current systems, but there were many voltages, two-phase and three-phase, 25 cycles, 35 cycles, 50 cycles, 60 cycles, and many types of equipment that were often incompatible with other equipment. With the passage of time direct current has almost disappeared; 60 cycles has become standard in the United States; voltages are generally multiples of 115, although 120 volts has become more common for residential service; and manufacturers have adopted standards that make most types of equipment interchangeable. Similar changes have taken place in most other countries, except that in some European countries 50 cycles is the standard frequency, and 220 volts is in common use on lighting circuits.

13

Electric Traction

PUBLIC TRANSPORTATION

The means of public transportation available before the coming of the steam railroad were saddle horses, carts, wagons, coaches, barges, sailboats, and steamboats. Overland transportation was limited until canals were constructed to link up with natural waterways.

RAILS AND RAILWAYS

For some purposes it was found to be desirable to provide rails made of timber or stone to guide the wheels of wagons or carts. Some used flanged wheels riding on rails consisting of iron straps screwed to wooden stringers, and later iron rails attached to wooden ties were used. Installations of this sort became common around mines, quarries, and sawmills. The cars were sometimes drawn by horses, sometimes pulled by cables, and in other cases pushed by men.

When the construction of the Liverpool and Manchester Railway had been nearly completed in 1829, the directors were undecided as to whether the motive power was to be the horse, the steam-driven cable, or the steam locomotive. George Stephenson finally proved that his steam locomotive was the most satisfactory method of propulsion. The steam locomotive proved its value, and from that time forward steam railroads were built in ever-increasing numbers.

STREET RAILWAYS

For urban transportation the horse-drawn coach or omnibus was widely used in larger cities and for interurban journeys. Construction of the first street railway, or tramway, was begun in New York City in 1832. The cars were horse-drawn, provided with flanged wheels, rolling on iron straps screwed to wooden stringers. Operation of this line was begun in 1833, but the project soon failed due to lack of patronage. A similar project was undertaken in Boston in 1836. A second attempt was made in New York City in 1852, using grooved iron rails laid on timber stringers. This form of rail proved to be troublesome because it caught the carriage wheels. In Philadelphia, a step rail was used on a street railway in 1855 with somewhat greater success. The first British tramway was built at Birkenhead in 1860. Between 1860 and 1890, horse-drawn street railways were constructed in all of the larger cities of the United States, and a somewhat similar development occurred in Europe but at a slower rate, due, in part, to the many narrow, winding streets.

The first elevated railroad began operation in New York City in 1867. Motive power at first was supplied by a cable, but a short time later steam locomotives were substituted. This project, as was the case with the first surface line, became involved in financial troubles but was reorganized and continued to operate. Other elevated lines were built in New York City, and eventually all of these were consolidated into one system.

In some instances the steam locomotives built for use on the elevated lines were used on surface lines. These were not the same as those in use on main line railways but were lighter and resembled the cars in the train that they pulled. Such applications of steam locomotives were confined largely to interurban lines, where greater speed was required.

ELECTRIC PROPULSION

Horsecars, for a time, seemed to solve the public transportation problem, but they were slow; and as the cities grew in size the need for faster transportation, or "rapid transit," became more urgent. The solution was found in the electrification of the street railways. The idea of electric propulsion was by no means new. Thomas Davenport, of Brandon, Vermont, whose name was mentioned previously in connection with his electric motor, built a small electric car which he operated on a circular track at Springfield, Massachusetts, in 1835. Robert Davidson of Scotland, in 1838, built a five-ton car propelled by an electric motor of his own design.

In 1847 Moses G. Farmer, whose name is already familiar because of his early experiments in electric lighting, constructed and operated an electric car capable of carrying two passengers. Professor C. G. Page of Washington made a trial run of a small electric locomotive on the Baltimore and Ohio Railroad, between Washington and Bladensburg in 1851. The motor was of the walking-beam type.

These and other experiments with electrically operated cars were interesting, but none of them was practical because the motors were inefficient and the batteries used for power were much too expensive. After the dynamo had been brought to a reasonable state of perfection and had been found to be reversible, electric power for traction and other purposes took on an entirely different aspect. At the industrial exposition in Berlin in 1879, Dr. Werner Siemens exhibited and operated an electric locomotive and cars over a circular track about 900 feet long, with power supplied by a dynamo. The motors on the locomotives for the first time were commutator-type machines, which were capable of a steady and powerful pull. Current was supplied through a third rail.

Two years later, in 1881, Dr. Siemens built an electric railway at Lichterfelde, a suburb of Berlin, as a regular commercial venture. This electric railway went into operation in the spring of 1881 and continued to carry passengers regularly thereafter.

Following Siemen's demonstration, numerous similar railways were built. Applications for electric railway patents filed in the United States Patent Office by Thomas Edison and Dr. Werner Siemens in 1880 were denied because of a caveat filed by Stephen D. Field in 1879. The Field and Edison interests were, however, pooled, and a company called the Electric Railway Company of the United States was formed. Edison built a small experimental electric railway at Menlo Park in 1880 and another in 1882. Field also built an experimental electric railway at Stockbridge, Massachusetts.

At the Chicago Railway Exposition held in 1883, the Electric Railway Company of the United States exhibited an electric railway. A third rail was used for this project, as was the case in the earlier Siemens exhibit, and the rail joints were bonded with heavy copper wire. Both the motor and the dynamo had been manufactured by the Weston Company.

ELECTRIFICATION OF STREET RAILWAYS

These demonstrations had shown that electric traction could become successful, and as a result street railway companies became deeply interested in electrification. There were many objections to horsecars and cable cars, the most important of which was that they were too slow. Cities were spreading out, and there was great need for moving large numbers of people from their homes to their places of employment.

In the United States those who became most active in this field, after Edison and Field, were Frank J. Sprague, Charles J. Van Depoele, Edward M. Bentley, Walter H.

Knight, Leo Daft, J. C. Henry, and Dr. Joseph R. Finney.

Sprague had joined Edison's staff during the period when Edison was installing electric lighting systems in the eastern states. He was a university man with a knowledge of the mathematics of electrical circuits and had been very valuable to Edison. After the installation of the Edison system at Brockton, Massachusetts, Sprague was put in charge. He continued his work with Edison but also devoted considerable time to his own research on the electric motor. He succeeded in designing an excellent motor and at the same time began to take a serious interest in electric railways. In April 1883 he resigned from Edison's organization and in November 1884 organized the Sprague Electric Railway and Motor Company.

Van Depoele, who was a skilled cabinetmaker and woodcarver, came to Chicago from Belgium in 1869. He built up a successful business and with his surplus funds began to experiment with electrical equipment. Within ten years he was installing arc lighting systems of his own manufacture, and in 1880 the Van Depoele Electric Manufacturing Company was established in Chicago. At about this time he became interested in electric railways, and in 1884 exhibited an electric railway in Toronto, which was equipped with overhead trolley wires. He received a United States patent covering this system a short time later, over the protests of Sprague, who had also invented overhead trolleys but was out of the country at the time.

Leo Daft installed and operated the Saratoga and Mount McGregor Railroad in 1883, using electricity for motive power. Later he electrified one of the New York elevated lines.

Knight and Bentley, who were employed by the Brush Electric Company at Cleveland, developed an electric traction system that operated successfully. Their first experimental car was put in service on one of the Cleveland lines on July 26, 1884. Although the car was kept in operation, the difficulties encountered were enormous. They experienced troubles with the motor suspension, the main drive to the axles, the central station generators, the motors, and other equipment. Valuable experience was gathered, however, and one by one the problems were solved. They abandoned the Cleveland experiment in 1887 and moved to Woonsocket, Rhode Island, where they had a contract with the street railway company for the electrification of the system. Equipment for this project was supplied by the Thomson-Houston Company. The first car on the Woonsocket lines was put in service in October 1887.

Knight and Bentley entered into another contract with the Observatory Hill Passenger Railway Company of Allegheny City, Pennsylvania. This project involved problems that had not been encountered previously, such as a 12 percent grade on a sharp curve. The work was completed in spite of the difficulties, and the first car was placed in service in January 1888. The problems that followed, for a time, seemed almost insurmountable. The motor brushes were laminated copper, and sparking was so severe that brushes and commutators were sputtering and burning up in rapid succession, leaving a trail of copper along the entire route.

The Sprague Electric Company undertook the electrification of the Richmond, Virginia, street railway system in 1887. After many construction difficulties, and after a case of typhoid fever, Sprague succeeded in getting portions of the electrified system into operation late that year. The Richmond undertaking was the most ambitious so far, in that it involved more miles of track, more cars, and many grades, some of which were steep.

THE CARBON BRUSH

Sprague encountered much the same difficulties as Knight and Bentley, who were still struggling to keep the cars moving at Allegheny City. The solution to the most vexing problem came toward the end of 1888, when Charles Van Depoele, who was then associated with the Thomson-Houston Company, recalled that he had once tried a carbon brush on a small motor and it had appeared to work satisfactorily. No one believed that a carbon brush with its high resistance could overcome the difficulties in railway motors, but the situation was so desperate that anything was worth a trial. Carbon brushes were tried and they worked. Not only did the brushes no longer burn up, but the com-

Figure 13.1 Sprague Electric Car *(Courtesy General Electric Company)*

mutators took on a polished, dark brown appearance. When carbon brushes were substituted for copper and brass on the railway motors, peace and tranquillity settled down upon the sorely harassed railway shops.

RAPID CONVERSION FROM HORSECARS TO ELECTRIC PROPULSION

From that point forward electrification of the street railways of the United States proceeded at a pace that taxed the ability of the manufacturers to supply the needed equipment. There was, of course, a steady evolution in motors, cars, control equipment, rails, roadbeds, overhead conductors, and power supplies. Among the important developments were the series-parallel controller, interpole motors, alternating-current distribution systems with rotary-converter or motor-generator substations, heavier rails, new methods of bonding, and higher speeds.

SUBURBAN AND MAIN LINE ELECTRIFICATION

The electric street railways developed into interurban lines, elevated railways, subway lines, and main line railways. Mention should be made also of trackless trolleys, which were buses equipped with electric motors fed from a double overhead trolley wire through a swiveled trolley pole.

THE DECLINE OF ELECTRIC STREET RAILWAYS

There are very few electric street railway systems remaining in the United States because of the lower cost and greater flexibility of buses powered by internal combustion engines. In Europe, however, where the automobile has not made such great inroads on the numbers of passengers carried, electric street railways are still operating successfully.

14

Electromagnetic Waves, Radio, Facsimile, and Television

A CENTURY OF PROGRESS

In the one hundred years between the closing decade of the eighteenth century and the late nineteenth century, electricity had progressed from a little-known, mysterious, natural phenomenon to a well-established science of great practical value. It had moved from the laboratory and lecture room into industry and commerce, where it touched the lives of all. The first giant step was taken by Volta, followed by Davy, Oersted, Faraday, Henry, Pixii, Berzelius, Ohm, Weber and Gauss, Ampère, Coulomb, Morse, Wheatstone, Swan, Edison, Bell, Maxwell, Helmholtz, and a host of others. The exploration of nature's great house had indeed progressed at an amazing rate, but there were many doors that had not yet been unlocked. What lay beyond one of these had been guessed by Faraday, Henry, and Maxwell, but the key was found by a young German professor named Heinrich Hertz (1857-1894).

HERTZ DISCOVERS ELECTROMAGNETIC WAVES

Hertz, as a young man, had been an assistant to the great Helmholtz at Berlin. The researches of Helmholtz and the mathematical explorations of Maxwell had convinced him that certain electrical phenomena are accompanied by electromagnetic waves in space. In 1887, at the age of thirty, when Hertz was professor of physics at the Technische Hochschule in Karlsruhe, he succeeded in proving the existence of such waves by means of rather simple apparatus. He set up a circuit containing an induction coil, a wire loop, and a spark gap. At the other end of the table he had a similar circuit containing only a spark gap. A discharge from the induction coil across the first gap was accompanied by a weaker spark across the gap in the receiving circuit. In order to prove that this experiment did indeed demonstrate the existence of electrical waves, he carried on further experiments for several years, and succeeded in reflecting and refracting these waves and in measuring their wavelength. Joseph Henry, many years earlier had nearly succeeded in proving what Hertz had found but had fallen a little short. He had even spoken of the phenomenon as resembling waves. Faraday had demonstrated that a magnetic field rotated the plane of polarization of light waves, and Hertz proved that electromagnetic waves were not limited to the very short ones. Maxwell's mathematical genius had been vindicated.

SIGNALING WITHOUT WIRES

As has been so often the case, great discoveries have been anticipated by others. In Galvani's first frog experiment an electrical impulse had been transmitted from an electrical machine to a scalpel without wires. Morse performed an experiment in 1842, in which he buried two plates in the ground on one side of a canal, connected through a battery and a key, and on the other side he buried two similar plates, connected through a galvanometer. He succeeded in transmitting signals by means of this equipment, but the results here were due merely to conduction.

Using somewhat similar equipment, Alexander Graham Bell performed experiments of the same kind on boats on the Potomac River in 1882 and succeeded in sending signals up to 1½ miles.

Edison, working with Gilliland, Phelps, and Smith in 1885, was able to establish communication between a moving train and stations along the way. The transmitter on the train consisted of a battery, key, and induction coil. The signal was picked up by a wire on poles along the right-of-way and was received as buzzing dots and dashes through a telephone receiver at the station. This system was installed on the Lehigh Valley Railroad in 1887 and operated successfully. The results were regarded at the time as merely the effects of induction, but the system certainly contained some of the elements of radio telegraphy.

GUGLIELMO MARCONI

Hertz and his colleagues were not greatly impressed with the importance of the discovery of electromagnetic waves, but others began to see the possibility that there could be practical applications for this newly discovered phenomenon. Near Bologna, Italy, a young man only twenty years of age, named Guglielmo Marconi (1874–1937), began in 1894 working with apparatus similar to that used by Hertz, with the definite purpose of using electromagnetic waves for communication without wires.

Marconi had one great advantage over Hertz because of the invention in 1890 by Branly of Paris of a device called a coherer. Branly had discovered that when certain metal filings were subjected to high-frequency alternating currents their direct current conductivity was thereby increased. Marconi found that the coherer could be used as a detector for electromagnetic waves. He was able to improve upon Branly's original invention by using silver plugs in a glass tube with a small gap between them which was filled with a mixture of silver and nickel filings.

Marconi was not alone in his search for a method of wireless communication. A Russian named Alexander S. Popov described his wireless telegraph apparatus before a gathering of the Russian Physical Society in St. Petersburg in 1895. His equipment was very similar to that which Marconi was using and included a coherer as a detector. Popov claimed that he had sent a wireless signal a distance of 600 yards. Because of Popov's address before the Physical Society the Russians have claimed priority over Marconi. Whatever the merit of these claims it was Marconi who developed the wireless telegraph into a practical reality.

FIRST RADIO PATENT

During the year 1895 Marconi was able to transmit wireless messages over a distance of a mile. He received a British patent in 1896. In 1897 a group of wealthy Englishmen, at the suggestion of Sir William Preece, head of the British Post Office, formed a company called the Wireless Telegraph and Signal Company, which gave Marconi about £60,000 in stock and £15,000 in money for the use of his invention and retained him to carry on further experiments. By the year 1900 wireless signals were transmitted a distance of 200 miles, and on December 12, 1901, Marconi received a wireless signal at St. John's, Newfoundland, which had originated at Poldu in Cornwall, England.

Many improvements were introduced as the result of years of experimentation, including improved antennas and far more sensitive detectors. Marconi's first detector was the coherer. For his transatlantic transmission in 1901, he used the mercury detector that had been invented a short time before by Lt. Solari of the Italian navy. Various forms of detectors had been invented over the brief early period of wireless telegraphy, including those already mentioned plus the electrolytic, the magnetic, and the hot-wire.

TUNED CIRCUITS

Maxwell's and Helmholtz's mathematical equations recognized the existence of electrical resonance, but it seems that Hertz, Marconi, and others were slow to realize its implications in practical electric circuits, especially in circuits involving high frequencies. Joseph Henry may have had some inkling of the need for tuning two circuits to respond to the same frequency. Multiplex telegraphs had made use of tuning at lower frequencies in order to separate the different messages coming over the same line. When Marconi applied this principle to wireless telegraphy, he achieved success. He was granted a patent on tuned wireless circuits that for many years was the basic patent in that field.

CONTINUOUS WAVES

The spark used by Marconi for his wireless telegraph produced a series of damped waves on a broad spectrum. Such a signal would be strongest in the region of the natural frequency of the transmitter circuit but would include many strong harmonics on both sides of the fundamental or natural frequency of the circuit. Each spark caused a group of electromagnetic waves of different frequencies that decayed to zero very quickly followed by another burst and another, at a rate determined by the interrupter on the coil. The effect on the receiver was a tone determined by the frequency of the interrupter. This tone was broken into dots and dashes by the key of the sender.

As early as 1892 Professor Elihu Thomson had discovered that a direct-current arc could produce undamped electrical oscillations. W. B. Duddell of England discovered in 1900 what was called the singing arc. Valdemar Poulsen of Denmark applied the discoveries of Thomson and Duddell to produce high-frequency oscillations for use in radio transmitters for which he received a patent in 1903. Fessenden used the Poulsen arc a short time later for wireless transmission. Despite the development of the Poulsen arc, most wireless stations were still using spark transmitters, but they had been greatly improved in the meantime so that interference was diminished considerably.

Poulsen formed his own company in 1910, and in the same year the Federal Telegraph Company bought his rights for the United States. The United States Navy adopted the Poulsen arc transmitter in 1912 or 1913 and used it until the end of World War I. The German Telefunken began using a form of arc transmitter as early as 1906, and Germany was therefore further advanced in wireless transmission than Britain or the United States.

DETECTORS

The principal forms of detectors in the early period of radio development were coherer, mercury, magnetic, hot-wire, electrolytic, and crystal. The coherer and mercury detector were mentioned earlier in connection with Marconi's work. It is rather surprising that a number of kinds of detectors had been invented before there was any serious effort at wireless communication. An early form of magnetic detector was devised by Ernest Rutherford in 1895. Marconi made two improved forms of magnetic detector in 1902. Altogether there were about a dozen forms of this type of detector.

R. A. Fessenden of the United States invented the electrolytic detector in 1902, and W. Schloemilch of Germany discovered the same principle a little later. The hot-wire detector goes back to the year 1890 when Rubens and Ritter made such a device in the form of a Wheatstone bridge. Fessenden received a U. S. patent on an improved and simpler form in 1902. There were other variations of such detectors that used the thermocouple principle.

G. W. Pickard of the United States began experimenting with crystals as detectors in 1903 and in 1906 found that silicon was an excellent material for that purpose. In 1906 also, General Dunwoody of the United States Army found that a combination of carborundum and carbon made a very stable detector. Many other crystals were found to be good detectors especially galena, which is lead sulfide. After World War I, and especially after broadcasting was begun, amateurs built crystal receiving sets in great numbers.

THE EDISON EFFECT

Edison had noted the blackening of his light bulbs in the plane of the filament and particularly near the positive terminal. In trying to find the reason for this phenomenon he had provided one of his lamps with a small plate inside the bulb near the filament that was connected to a lead-in wire. He found that a galvanometer connected to this plate and to the filament showed a small current, when the lamp was in operation, and that the plate was negative with respect to the filament. He discovered also that when a battery was connected to the filament and to the plate, a current could be made to flow in one direction but not in the other. He noted also that when the battery voltage was increased sufficiently, a blue glow appeared around the filament. The phenomenon, relating to the conductivity of the evacuated space in the lamp, called the Edison effect, was discovered in 1883 and was patented by Edison.

Elster and Geitel in Germany began an investigation in 1882 of ionization caused by incandescent metals. About the year 1888 they carried out experiments with evacuated bulbs and rediscovered the Edison effect. Hittorf found in

1884 that the space between hot and cold carbon electrodes was partially conducting.

THE FLEMING VALVE

William H. Preece, who was at the head of the British telegraph service, about 1888 conducted a series of experiments with an Edison bulb containing a plate, in which he verified Edison's discovery. Similar experiments by J. A. Fleming about the year 1890 yielded the same results, concerning which he read a paper before the Royal Society in the same year and another in 1896.

It was during this period that the discoveries of Mme Curie, J. J. Thomson, William Crookes, Rutherford, Lenard, Planck, and others were being made. J. J. Thomson in 1897 discovered and identified the electron, which he called a corpuscle. These discoveries served to clarify some of the mysteries surrounding the Edison effect, the Crookes tube, ionization of liquids, and the spontaneous emissions from certain elements.

Fleming's interest in the Edison effect was revived as a result of the new discoveries and because of the Marconi experiments. Realizing that what was needed for the wireless telegraph was a more efficient detector, he set about the task of adapting the Edison bulb to this purpose. He made bulbs in which a part of the filament was surrounded by a metal cylinder connected to a separate terminal. This device, called the Fleming valve, served as a detector and rectifier, but as a detector it was very little better than a crystal. The Fleming valve was brought out in 1904 and was covered by a patent.

DE FOREST AUDION

Lee De Forest of the United States, who had been investigating the conduction of electricity through gases since

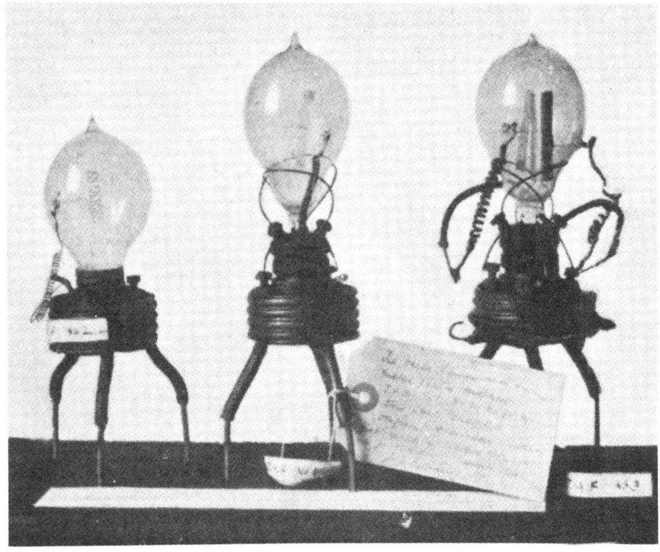

Figure 14.1 Fleming Oscillation Valves *(Courtesy Burndy Library)*

1900, was also working on the Edison effect and wireless detectors. He filed patent applications for two-element tubes in November 1904 and in January, February, and April 1905. These tubes had two plates or wings. On October 20, 1906, before a meeting of the American Institute of Electrical Engineers held in New York, De Forest made his first announcement of the invention of the three-element tube containing a grid. He called his triode tube the "audion," but he apparently had no notion at the time of what a giant this new tube was to become. To him it was merely a wireless detector, by far the most efficient so far discovered.

AMPLIFICATION

In 1911 Lieben, Reisz, and Strauss of Austria were granted a French patent covering the use of triode tubes as amplifiers. These were connected through transformers so that the plate of the first tube was connected to the grid of the second, and so on.

ARMSTRONG'S OSCILLATOR TUBE

In the year 1912 a young man twenty-two years of age, then a student at Columbia University, made a discovery that again completely transformed wireless communication and made possible most of the progress which followed. The young man was E. H. Armstrong, and the discovery was the fact that when suitable connections were made from the grid of a tube to its plate the tube would oscillate and would generate high-frequency alternating currents that could be used as the source of continuous electric waves.

De Forest made the same discovery at about the same time, but Armstrong was given priority, although his patent was not filed until October 29, 1913. Besides Armstrong and De Forest, it appears that Round and Franklin in England and Meissner in Germany had approached very closely to the same discovery at about the same time.

The European war began in August 1914, and radio became a matter of great military importance. New developments in radio came very rapidly but were often cloaked in great military secrecy. The radio telephone soon came into use for ship to ship, and ship to shore communication, but for long-distance work radio telegraph was still used almost exclusively. For such long-distance communication the sources of electric waves were still the spark for damped waves, and the Poulsen arc for continuous waves.

THE ALEXANDERSON HIGH-FREQUENCY GENERATOR

R. A. Fessenden had made an attempt in the early 1900s to build a high-frequency generator. He took his idea to the

Figure 14.2 Ernst F. W. Alexanderson with His High-Frequency Generator. This is the later model of his alternator used in radio stations after 1918. *(Courtesy General Electric Company)*

General Electric Company, where Ernst F. W. Alexanderson was asked to find a solution. After working on the problem for two years, he produced a small generator which could operate at 60,000 cycles, and a little later at 100,000 cycles. The first trials of the Alexanderson high-frequency generators were made in 1906, with somewhat indifferent results. Little was heard of high-frequency alternators until 1918 when a 200-kilowatt machine was installed at the American Marconi Company station at New Brunswick. Signals from this new station were strong and clear.

The Marconi Company offered to buy exclusive rights to the Alexanderson generator, but the General Electric Company was dissuaded from taking such a step. Finally, in 1919, a new company called the Radio Corporation of America was formed under the sponsorship of the General Electric Company to exploit the new device. Later the American Telephone and Telegraph Company, the Western Electric Company, the Westinghouse Electric and Manufacturing Company, and the United Fruit Company were invited to participate. In 1932 RCA became an independent corporation, and its stock was widely distributed.

AMATEUR RADIO AND RADIO BROADCASTING

When the European war ended in 1918, radio was back in private hands, and with the change came a wonderful new interest on the part of amateur radio operators. Many of these had been radio operators in the various navies or armies and had sufficient knowledge to build sending and receiving sets. These amateurs were licensed by the federal government and were assigned station call letters. At first their messages were in international Morse code, but later they changed to voice. Their work greatly accelerated the development of radio, especially shortwave. The amateurs had been assigned the higher frequency portion of the radio spectrum that was thought to have been less desirable, but they soon led the way in long-distance radio communication.

In 1920 the Westinghouse Electric and Manufacturing Company established the first broadcasting station in the world, station KDKA at Pittsburgh. Within a very short time there were hundreds of such stations throughout the United States and in Europe. The American stations, however, were privately owned and depended upon advertising for their support, whereas European stations were owned by the various governments and expanded more slowly.

The response of the public was immediate, and almost everyone that had any mechanical ability was building a crystal receiver equipped with headphones. A short time later they were building single-tube receivers with power supplied by a six-volt storage battery, and after another brief period came multitube radios with amplifiers and loudspeakers. The storage battery was messy and caused acid burns in clothing, rugs, and floors. As a substitute for batteries the manufacturers soon provided rectifiers called battery eliminators, but these were not entirely satisfactory. There was a great need for radio tubes that would operate on alternating current. This need was satisfied in 1926 by the invention of the alternating-current tube by H. M. Freeman and W. G. Wade of the Westinghouse Company.

The growth of the radio industry was phenomenal. By December 1922 there were 569 radio broadcasting stations in the United States and an estimated 1,500,000 to 2,500,000 receiving sets. The broadcast and other frequencies were soon becoming crowded so that it became necessary to establish frequency bands for various purposes.

Regulation of the radio broadcasting industry was essential, particularly in the matter of frequency allocation but also in the power allowed to any one station.

REGULATION OF RADIO

As early as 1910 Congress had passed the Wireless Ship Act, which set forth requirements for wireless on ships and regulated its use. An international conference on radiotelegraphy, held in Berlin in 1906, had issued certain regulations governing wireless that were agreed to by all of the important countries. In 1912 Congress passed the Radio Act, under which the secretaries of commerce and labor were made responsible for the licensing of radio stations and radio operators. Experimental broadcasting was approved in 1919 for limited commercial stations on a wavelength of 360 meters.

The first National Radio Conference, held in 1922, made certain recommendations as a result of which the secretary of commerce issued new regulations governing broadcasting stations. These regulations provided for a minimum power of 500 watts with a maximum of 1000 watts and permitted frequencies between 750 and 833 kilocycles. After the National Radio Conferences of 1923 and 1924, the Department of Commerce allocated the frequency band between 550 and 1500 kilocycles to radio broadcast and permitted power up to 5000 watts. Despite all attempts to cope with the tremendous increase in the number of radio broadcast stations, regulation was falling far behind the needs of the industry, so that interference between stations was widespread, especially after sunset.

The fourth National Radio Conference, called in 1925, asked for limitations on broadcast time and station power, but several court decisions held that the secretary of commerce had no power to impose such restrictions under the Radio Act of 1912. Because of these court decisions the entire broadcasting industry was in chaos. President Coolidge asked Congress to act in the matter, and the result was the Dill-White Act of 1927, which established the Federal Radio Commission. This commission had regulatory powers over radio, including the issuing of licenses, allocation of frequencies, and the control of station power. The act also empowered the secretary of commerce to inspect radio stations, to examine and license radio operators, and to assign station call letters.

FEDERAL COMMUNICATIONS COMMISSION

In 1933 the secretary of commerce appointed an interdepartmental committee to study and report on the national and international radio situation. The committee recommended that a commission be established with broad powers to regulate communications of all kinds. In 1934 the Congress passed an act establishing the Federal Communications Commission, or FCC as it is generally known. Although the commission was empowered to regulate all forms of communication, its most pressing problems were in the field of radio and television, which was just beginning. The new commission succeeded admirably in bringing order into what had long been a very confused situation. The discussion here has dealt only with the United States, but the problem was worldwide, and all other nations found it necessary to adopt regulations similar to those described for the United States. From time to time international agreements regulating radio were made particularly in the matter of assigned frequencies.

FREQUENCY ALLOCATIONS

At the present time the frequencies assigned to AM radio broadcast lie in the band between 535 and 1605 kilocycles, and for FM broadcast between 88 and 108 megacycles, with a total of 100 channels. Certain stations are permitted to broadcast during daylight hours only, whereas other large and important stations are assigned to clear channels and 24-hour service, with power up to 50,000 watts. Call letters have been established by international agreement, and for the United States broadcast stations east of the Mississippi, in general, have call letters beginning with W, and those west of the Mississippi beginning with K. FM stations carry the suffix FM after the call letters.

In addition to broadcast radio and TV the FCC has allocated frequency bands to a host of other applications including Safety Services (aviation, marine, police, fire, local government, forestry, highway, maintenance, special emergency, state National Guard, and disaster), Industrial Services (power, petroleum, forest products, business, manufacturers, relay press, motion picture, industrial radiolocation, and telephone maintenance), Land Transportation Services (railroad, motor carrier, taxicabs, and automobile emergency), Amateur Radio, and Citizens Radio. In addition to these there are such other frequency bands as Government Radio, Radar, Loran, Ramark, and Radio Relay Telephone.

RADIO RECEIVERS

The improvements in radio transmitters were accompanied by corresponding improvements in radio receiving sets and particularly in the quality of loudspeakers. In the early receivers headphones were used almost exclusively, and they were followed by loudspeakers that were essentially overgrown telephone receivers. These were very deficient in the higher tones and produced raucous sounds that were difficult to understand. This deficiency was soon recognized and resulted in the adoption of the speaker cone. There were also improvements in microphones or pickups used by the transmitting stations, so that by the middle 1920s the quality of reception had improved tremendously. A similar improvement occurred in phonograph recordings with the advent of electrical transcription and later with the change from wire to tape recorders.

In addition to the improvements in the sound system there were equally important changes in the electrical circuits of receiving sets. The early crystal and single-tube receivers used a regenerative circuit patented by Major E. H. Armstrong in 1914. The regenerative circuit was in effect a low-power transmitter that caused howls and squeals, due to heterodyning other receivers or transmitting stations. Heterodyning refers to the superposition of one signal upon another of slightly different frequency resulting in a beat frequency signal equal to the difference between the two. In 1919 Armstrong invented the superheterodyne circuit for which he received a United States patent on June 8, 1920. Some time elapsed before this circuit was fully appreciated but eventually it became the most widely used of any of the circuits. Another somewhat similar circuit, which was widely used for a time, was the tuned radio-frequency circuit invented by Louis Alton Hazeltine in 1924. Improvements and modifications of the superheterodyne circuit were made by Hartley, Colpitts, and Meissner.

At first the radio-frequency stages of receiving sets were tuned by separate knobs and required a great amount of patience as well as luck in tuning in a particular station. It was thought that it was impossible to construct the variable condensers and other components with sufficient accuracy to mount them on a single shaft, but this result was accomplished after a time with the aid of small trimmer condensers.

One of Armstrong's greatest accomplishments was his invention of the FM system of radio broadcast. In the FM system the frequency of the signal is modulated rather than the amplitude. FM is practically free from interference from static or other radio signals.

FACSIMILE TRANSMISSION

Before discussing television it will be helpful to trace the development of facsimile transmission, which, in some respects, was the forerunner of television. About the year 1840 Alexander Bain of Scotland invented a crude method of sending facsimile over telegraph wires that contained certain elements used in later facsimile and television, especially the scanning principle. In the Bain invention, the characters to be sent were first set up in metal and were connected to one side of a battery. A pendulum carrying a metal stylus was set to swinging over the metal characters so that the stylus made light contact with them. The other side of the battery and the pendulum were connected to the line. At the other end of the line was a similar pendulum and stylus, which traced the characters on a chemically treated paper. The pendulums served to synchronize the sending and receiving apparatus. Clockwork was used at both ends to advance the metal characters and the paper so as to scan successive lines. There was great difficulty, however, in maintaining synchronism, and because of the cumbersome methods involved the scheme had no practical value.

About a year later Frederick Bakewell, an Englishman, invented the prototype of most later facsimile apparatus, in which the message was wrapped around a revolving cylinder and was moved along at a uniform rate by a screw. The message was scratched through a coating of varnish or shellac covering a thin metal sheet and was picked up by a metal stylus. At the receiving end the message was transcribed on a sheet of chemically treated paper wrapped around a metal cylinder. Both cylinders were revolved by clockwork at the same speed, but synchronism was almost impossible.

COMMERCIAL FACSIMILE

Giovanni Caselli built an apparatus on much the same principles but with much greater mechanical refinement and operated his apparatus commercially over a telegraph line between Paris and Lyons for five years from 1865 to 1870. D'Alincourt made improvements upon the Caselli devices by using tuning forks for synchronizing. Further improvements were made about 1875 by Francis de Hondt of Chicago and William Sawyer of New York.

PHOTOELECTRIC DEVICES

Noah S. Amstutz of Cleveland demonstrated a facsimile apparatus in 1891 in which the transmitting device contained a selenium cell and the receiver reproduced the picture in halftone by means of dots. The photoelectric properties of selenium had been discovered by Willoughby Smith in 1873, and its applications were improved upon by Shelford Bidwell, both of whom were from England.

In the late years of the nineteenth century, Dr. Arthur Korn of Germany improved upon the Amstutz apparatus by introducing a selenium cell in the receiving set which acted as a relay or amplifier to reinforce the incoming signal and to overcome to some extent the sluggishness of the selenium cell in the transmitting device. This system was successfully demonstrated in 1902.

After the invention of the vacuum tube, it became possible to amplify the incoming signals even more. As a next step the selenium cell was replaced by the photoelectric cell, or photocell as it is now more commonly called. In the photocell various compounds of cesium, potassium, rubidium, tantalum, thorium, or zirconium were used as the light-sensitive material. These were deposited on a metal plate that was then enclosed in an evacuated glass tube with suitable connections to the outside. Some of the later tubes were gas-filled The new photocells responded much more quickly to changes in light intensity and covered a greater portion of the visible spectrum than had been the case with the older selenium cells. Another improvement of great importance to facsimile was the change to synchronization by means of alternating currents or radio signals.

PICTURES BY CABLE

H. G. Bartholomew and M. L. D. MacFarlane of England invented an apparatus that made possible sending pictures by cable. The impulses from the photocell were first recorded in the form of perforations in a paper tape that was then fed into a standard telegraph transmitter. A similar tape was made at the receiving station, which could then be put into a machine that reproduced the picture.

By 1924 the Western Union Telegraph Company was furnishing regular picture service to newspapers. In the same year RCA sent pictures by radio to Europe.

TELEVISION

The greatest difference between facsimile and television lies in the speed at which pictures must be transmitted. The earliest attempt at transmitting visual images by wire was made in 1875 by G. R. Carey of Boston, who built an apparatus in which the sending station had a mosaic of small selenium cells, each of which was connected by wire to a small shutter in a corresponding mosaic at the receiving station. When an image was focused through a lens on the selenium mosaic, each cell responded according to the intensity of light falling upon it. The corresponding shutters at the receiving end were deflected accordingly and produced a crude reproduction of the original image.

THE SCANNING DISK AND MECHANICAL TELEVISION

Dr. Paul Nipkow of Berlin invented the first mechanical scanning disk in 1884, which was the forerunner of the various mechanical scanners used in television transmitters for many years thereafter. In general, it consisted of a metal disk having a series of holes arranged in a short spiral. As the disk revolved at rather high speed, light impulses were received through the holes, scanning the various parts of the picture to be transmitted. These momentary light impulses fell upon a selenium cell that was excited in accordance with the intensity of the light falling upon it. The small and rapidly changing electrical impulses from the selenium cell caused a neon light at the receiving station to change its light output accordingly. The light from the neon bulb passed through the holes of a scanning disk like that at the transmitter, revolving in synchronism with it. The light coming through the holes in the scanner fell upon a ground-glass screen and produced an image of the original picture.

Television sets of this same general type with various improvements were in operation until 1933, but none of these was satisfactory or commercially useful. These mechanical television sets operated at twenty-four frames per second, with sixty lines per frame.

THE ICONOSCOPE

Modern television had its beginning with the invention of the iconoscope by Dr. V. K. Zworykin, who at that time was employed by the Westinghouse Company but who later joined the staff of RCA. Zworykin's patent was filed in the United States Patent Office on December 29, 1923, but was not granted until December 20, 1938. The iconoscope was the essential part of the television camera and was the first such device to scan a picture entirely by electronic methods. The receiving apparatus, called a kinescope and now generally called a picture tube, was a modification of the cathode-ray oscilloscope invented by Karl F. Braun of the University of Strasbourg in 1897. In 1923, shortly after his invention of the iconoscope, Zworykin set up a transmitting and receiving station, connected by wires, operating at 60 cycles, with which he was successful in transmitting television pictures.

On January 7, 1927, Philo Farnsworth filed a patent application covering an electronic camera tube, called the dissector tube, which in some respects was an improvement over Zworykin's iconoscope.

Although the essential parts and technique were at hand for electronic television, it had not yet been established that electronic television was superior to the older mechanical equipment. On April 7, 1927, the American Telephone and Telegraph Company demonstrated the transmission of mechanical television by wire, between New York and Washington. A few days later, on April 16, 1927, the American Telephone and Telegraph Company demonstrated for the first time the transmission of television signals by radio. This demonstration was still by means of mechanical television and took place between New York City and Whippany, New Jersey, about 27 miles distant.

IMPROVEMENTS ON THE ICONOSCOPE

Various improved forms of television tubes followed the iconoscope, beginning with the dissector tube of Philo Farnsworth in October 1934, the emitron by John Baird in November 1936, the orthicon by RCA in June 1939, followed by the image orthicon and the vidicon. There were similarities in all of these, but the later tubes were much more sensitive and covered a greater part of the light spectrum. All of them employed a stream of electrons from an electron gun to scan the image. This process of scanning is accomplished with remarkable accuracy by moving the electron stream from side to side and downward line by line by means of electrostatic and magnetic fields. In the earlier tubes, including the iconscope and dissector tube, the light from the scene to be televised and the electron stream impinged on the same photosensitive surface. In the later camera tubes the electrons are directed toward a metal surface in back of the light-sensitive mosaic. This mosaic screen is made up of a deposit of microscopic globules of a cesium-silver compound on a sheet of mica. Although the coating of tiny globules appears to the naked eye to be continuous, the individual globules are actually separated and insulated from one another. The cesium-silver compound has the property of emitting electrons when light falls upon it, and the quantity of electrons emitted varies with the light intensity. When electrons are lost, the globules become positively charged. In the iconoscope and dissector tube, there is a metal plate attached to the back of the mica that behaves like a condenser plate. As the electron beam moves back and forth across the cesium-silver mosaic, the positive charges on the globules are partly or wholly neutralized thereby, changing the potential of the back plate and causing it to give up a tiny alternating current of high frequency. This alternating current is then amplified and becomes the video signal.

Other types of television camera tubes operate in a similar fashion except that the electron stream strikes a plate on the back of the mosaic, changing its potential to correspond with the changes in potential of the globules in the mosaic. It must be remembered that each charged globule holds a bound charge of the opposite sign on the metal plate directly opposite its position, so that there is a mosaic of bound charges on the metal plate and not merely a charge on the plate as a whole.

In the image orthicon the electron beam is reflected back to an electron multiplier that serves to amplify the signal greatly. The image orthicon is about one hundred times as sensitive as the iconoscope and can therefore be used in ordinary or subdued light, whereas the iconoscope required intense illumination.

TRANSMISSION BY RADIO WAVES

In order to transmit the information from the television camera by radio, the carrier wave must be modulated to correspond with tiny impulses received by the light-sensitive mosaic. Transmission of the signal is complicated by the need for sending out another signal at the same time to carry the sound. Both signals are sent out by the same transmitter after mixing. They occupy the same frequency band, but the video portion is transmitted by amplitude modulation while the audio signal is frequency modulated.

In the receiving set the video signal, after being separated from the audio, enters the kinescope, which in some respects resembles the camera tube, and reproduces the picture on a fluorescent screen. Synchronism between the transmitter and receiver is maintained by means of a special signal at the end of each line and frame. Audio reception is of course exactly like ordinary FM reception except for the frequencies employed.

While ordinary black-and-white television was perhaps the most sophisticated mass-produced electronic apparatus devised up to the time of its inception, color television is far more complex. In general, color television uses combinations of red, green, and blue to synthesize approximations to the colors of the spectrum.

The camera for color television has three separate tubes, one each for the colors red, green, and blue. In the receiver there are three scanning guns, working in synchronism, which cause the red, green, and blue phosphors to glow in their characteristic colors.

REGULATION OF TELEVISION AND CHANNEL ALLOCATIONS

In 1941 the Federal Communications Commission established 525 lines per frame and 30 frames per second as the standard for the United States. Different standards exist in other countries. The FCC also allotted frequency bands in which the channels were separated by a minimum of 6 megacycles. In the United States the frequencies allotted to television are as shown in Table 14.1.

Table 14.1 Frequencies of Television Channels

VHF				UHF	
Channel	Megacycles	Channel	Megacycles	Channel	Megacycles
2	54	8	180	14	470
3	60	9	186	24	530
4	66	10	192	34	590
5	76	11	198	44	650
6	82	12	204	54	710
7	174	13	210	64	770
				74	830
				83	884

The initials VHF and UHF are used to designate very high frequency and ultra high frequency. For Channel 2 the wavelength is 5.56 meters, and for Channel 83 the wavelength is slightly more than $^1/_3$ meter.

15

The Crookes Tube, X Rays, Radioactivity, Structure of the Atom, Accelerators and Atomic Research

THE CROOKES TUBE

No other invention or discovery in the electrical field has had so many important and useful descendants as the Crookes tube. From it, directly or indirectly, came the X-ray tube, the cathode-ray tube or oscilloscope, the radio vacuum tube, the iconoscope and its successors, the kinescope, a number of accelerators, the electron gun, and the electron microscope. The Crookes tube might properly be called a cathode-ray tube, but that term has for some reason been applied to the Braun oscilloscope. The Crookes tube, in addition to its importance as the ancestor of so many other devices, led to the discovery of radioactivity.

Essentially the Crookes tube is an evacuated glass vessel containing a cathode or negative electrode and an anode or positive electrode. The construction of such a tube requires only a skilled glassblower and a good air pump.

Glassblowing is an ancient art, but the making of chemical and other scientific apparatus did not develop until the fourteenth or fifteenth centuries. The glassblowers of that period were hampered because of the lack of gas or liquid fuels. By the nineteenth century glassblowing was far advanced, as exemplified by the work of Geissler.

Von Guericke produced the first air pump in 1654; constructed of glass, it resembled a syringe. Similar pumps by Boyle, Hawksbee, and Smeaton had improved efficiencies. Later, still more efficient pumps, such as the oil-air pump, were made of metal. In the oil-air pump the space below the piston was filled with oil which served to eliminate the air that would otherwise be trapped.

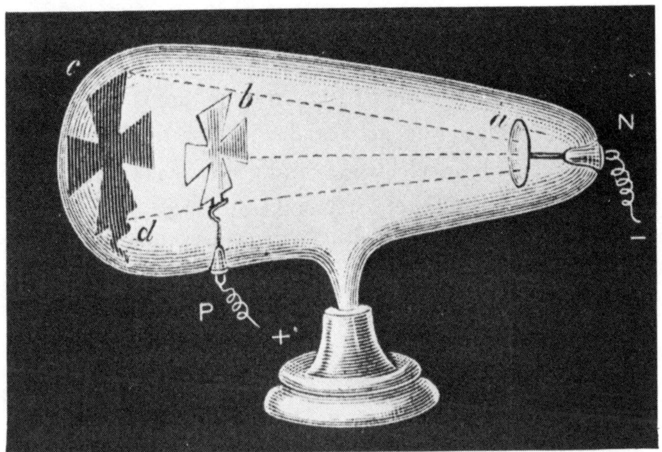

Figure 15.1 A Diagram of the Crookes Tube *(Courtesy Burndy Library)*

The first mercury pump was made by Swedenborg in 1722, followed by mercury pumps by Baader in 1784, Hindenberg in 1787, Edelkrantz in 1804, and Patten in 1824. Geissler's mercury pump of 1855 was capable of producing a vacuum of 0.05 millimeters of mercury. Töpler improved the Geissler pump somewhat, but the Sprengel pump of 1865 was a great improvement in that it reduced the labor and attention required and speeded up the rate of evacuation. The Gaede pump of 1907 was a rotary pump, mechanically operated, which produced a vacuum of 0.00004 millimeters.

VACUUM TUBES BEFORE CROOKES

The earliest experiments in vacuum tube phenomena were those of Picard and Hawksbee. Picard observed flashes of light in the space above the mercury in a Torricellian tube when the tube was shaken, and Hawksbee noted the fluorescence in an evacuated glass globe when it was electrified.

These experiments differed from those with the Crookes tube because no internal electrodes were involved.

The first vacuum tube of the Crookes variety apparently was constructed in Michael Faraday's laboratory in 1838. It contained two brass electrodes and was evacuated to a pressure of about 2 millimeters. When a discharge from an electrical machine was sent through the tube, Faraday observed a purple glow near the anode and a dark space nearer the cathode. This dark space has since been called the Faraday dark space.

Masson of France, who was one of the early induction coil makers, constructed a vacuum tube with two electrodes in 1853 and discovered that when a discharge from an induction coil was sent through the partially evacuated tube a bright pink glow filled it. This phenomenon occurs with a pressure of about 5 millimeters of mercury. Theoretically the discharge from an induction coil should be a pulsating alternating current that would not be well suited for use in a vacuum tube. Actually, however, the secondary voltage is so much higher when the circuit is broken by the interrupter than when the circuit is closed that the secondary output is almost unidirectional.

Heinrich Geissler of Bonn made a great number of vacuum tubes that were shipped to schools, colleges, and other laboratories everywhere, where they were used for instruction and entertainment. Most of these contained air at about 5 millimeters of mercury and exhibited the characteristic pink glow for that pressure. Others contained gases other than air, and some were made of glass containing metallic salts that fluoresced in various colors.

Julius Plücker, also of Bonn, discovered in 1858 and 1859 that as the vacuum is increased the Faraday dark space becomes larger, and he noted the greenish phosphorescence of the glass. He also observed the effect of a magnetic field upon the discharge. J. W. Hittorf, one of Plücker's students, found in 1869 that an opaque object placed in front of the cathode cast a shadow upon the glass.

Eugen Goldstein of Germany performed a long series of experiments with vacuum tubes of many shapes, having a variety of electrodes. He found that the cathode rays are emitted perpendicularly to the surface of the cathode, so that a concave cathode would tend to focus the rays, but the sharpness of the focus was influenced by the residual pressure and the potential difference. He found that the properties of the rays were independent of the material used in the cathode, and he found also that the rays could produce chemical reactions.

Cromwell Varley advanced the idea in 1871 that the cathode rays were negatively charged particles but with only partial proof.

SIR WILLIAM CROOKES AND HIS EXPERIMENTS

We have very little information about Crookes's early experiments with the vacuum tube that now bears his name, but it appears that he must have begun his work during the 1860s, because it was reported that he demonstrated his paddle wheel experiment in 1870. In this experiment he provided two rails inside the tube on which the axle of the paddle wheel could roll. In this experiment the paddle wheel, which had four vanes, was made to turn by the bombardment of what Crookes later called molecular light or radiant matter, emanating from the cathode.

Crooke's later tubes were constructed by a skilled instrument maker named Gimingham and were undoubtedly the finest so far produced. With them Crookes carried out many experiments that were published in several papers

Figure 15.2 Sir William Crookes *(From Smithsonian Institution)*

delivered in 1878 and 1879. In addition to the facts already known, Crookes found that the character of the discharge changed as the degree of exhaustion increased. The first discharge occurred in the form of bluish streamers when the pressure reached about 10 millimeters of mercury. At about 5 millimeters the tube was completely filled by a pink glow. At about 2 millimeters the Faraday dark space appeared near the anode, and there was a bluish negative glow near the cathode. At successively lower pressure steps, striations appeared in the vicinity of the anode stretching out toward the cathode, and another dark space, now called the Crookes dark space, appeared surrounding the cathode. The dark space was in turn surrounded by the blue negative glow.

As the pressure was further reduced, the Crookes dark space increased in size until finally it encompassed the entire tube. Crookes speculated that the dark space near the cathode was a region in which the cathode rays had a free path before colliding with gas molecules. The blue negative glow was caused by these collisions. This theory was later confirmed, and it was found that the cathode-ray particles, later called electrons, had enormous velocities of some tens of thousands of miles per second as they were shot out from the cathode.

Other inventions or discoveries by Sir William Crookes included the radiometer in 1875 and the spinthariscope in the early 1900s. He also observed the fluorescence of various minerals under the influence of ultraviolet light from his tube. Crookes was above all a chemist of considerable stature and was the discoverer of the element thallium.

LATER DEVELOPMENTS IN CATHODE RAYS

Jean Perrin of Paris in 1875 confirmed Varley's hypothesis that cathode rays were negatively charged particles. Heinrich Hertz and his assistant P. Lenard succeeded in passing cathode rays through an aluminum window at the end of the tube opposite the cathode and found that the rays were capable of producing fluorescence outside the tube.

J. J. Thomson in 1897 found the ratio of the charge *e* to the mass *m* of the corpuscle, as he called the electron, and he found also that after leaving the cathode these particles travel at a speed of about one-fifth of the velocity of light, depending upon the applied voltage and the degree of exhaustion of the tube. In 1906, Millikan, by means of his oil drop experiments, was able to determine the value of *e* as -1.609×10^{-19} coulombs. When this value was substituted in Thomson's value for e/m Millikan found the mass of the electron to be 9.1072×10^{-28} gram.

The term electron was first suggested by Dr. G. Johnstone Stoney as a name for the natural unit of electricity in 1891. Later he applied the term to the cathode-ray particles that J. J. Thomson called corpuscles. It was not until the year 1900, or a little later, that the name came into general use.

X RAYS

Among the many descendants of the Crookes tube was the X-ray tube. On November 8, 1895, Wilhelm Conrad Röntgen, a professor of physics at Würzburg University, was experimenting with a Crookes tube when he discovered that a screen coated with a thin layer of barium platinocyanide lying nearby fluoresced when the tube was operating. He found that objects placed between the tube and the screen cast shadows upon the screen that were darker for objects of greater density. When the hand was interposed, the bones cast darker shadows than the flesh.

The source of the rays coming from the tube, which Röntgen called X rays because their nature was unknown to him,

Figure 15.3 Wilhelm C. Röntgen *(From Smithsonian Institution)*

was found to be the inside surface of the glass tube and the anode upon which the cathode rays impinged.

For several years after Röntgen's discovery, there was a difference of opinion among scientists as to whether X rays were particles or very short electromagnetic waves. It became clear after a time that they were probably electromagnetic waves, but the matter was not definitely settled until about 1912, when Max von Laue of Berlin with his associates Friedrich and Knipping performed the crucial experiment in which X rays in a small beam were allowed to fall upon a thin crystal where they were diffracted. The diffracted rays falling upon a photographic plate exhibited symmetrical patterns about a dark central spot.

Von Laue's experiment led W. H. Bragg and his son W. L. Bragg of London University to extend the experiment by using a crystal lattice to obtain an X-ray spectrogram. Instead of sending the X rays through a small hole in a lead plate to obtain a fine beam, they used slits; and instead of sending the X rays through a crystal, they reflected them from the face of a crystal. The crystal was mounted so that it could be rotated about an axis at the point at which the X rays were incident upon the crystal face. A strip of photographic film was arranged in an arc of a circle whose center was the axis upon which the crystal rotated. The spectrogram obtained by this method was similar to that made by a spectroscope with ordinary light. The lines on the spectrogram were due to X rays of different wavelengths reinforced by reflections from deeper layers of the crystal lattice. Since the thickness of the layers could be determined, it became possible to find the wavelengths of the X rays. Even more important was the fact that an arrangement of this kind made it possible to study the structure of crystals.

Many improvements were made in X-ray tubes, some of which came soon after Röntgen's discovery, but the most important improvement was made by William D. Coolidge of the General Electric Company in 1913, when he substituted a hot filament cathode for the cold cathode used previously. The Coolidge X-ray tube provided many more electrons than the older tubes and was much more efficient. The source of electrons in the cold cathode tubes was principally the small amount of residual gas. Since the Coolidge X-ray tube had no need of this gas it could be evacuated to the highest degree possible.

X rays were used in the Vienna hospitals within a few weeks after Röntgen's discovery and were soon in use for medical purposes everywhere. Many other uses were found such as metallography, crystallography, detection of hidden objects, and a very important use in the advancement of research in atomic physics.

RADIOACTIVITY

After the discovery of X rays, there was great interest in the possibility that other still unknown kinds of radiation might exist. Scientists were spurred on also by Hertz's discovery of radio waves and by the particles known to exist in a Crookes-tube stream. Henri Becquerel of Paris tried wrapping various minerals in black paper and placing them on photographic plates to determine whether or not there was natural radiation from such minerals. He found that uranium compounds did cause a blackening of the plates. On February 24, 1896, he delivered a lecture on the subject before the French Académie des Sciences in which he mentioned the facts that, not only the compounds of uranium, but the metal itself caused the plates to darken and that the radiation, whatever its nature, would discharge either a positively or negatively charged electroscope.

Within a short time G. C. Schmidt and Mme Marie Curie discovered that thorium produced effects similar to uranium and that other elements showed slight radioactivity.

In Bohemia, at Joachimsthal, there was a mining operation in which pitchblende was extracted, from which uranium was recovered. Mme Curie had found that the residue, after the extraction of uranium, was radioactive and arranged with the Austrian government to obtain about a ton of the material. She and her husband Pierre set to work on a long chemical process of refining and re-refining the pitchblende to produce an increasing concentration of the radioactive substance. In 1898 they discovered the element polonium (atomic weight 210) and five months later discovered radium (atomic weight 226). Polonium, named after Mme Curie's native country, is highly radioactive but has a half-life of only 140 days. Radium, on the other hand, has a half-life of 1600 years. Various other radioactive substances were found within a short time including radon and actinium. Radon was an emanation from radium but was found to be an element with an atomic weight of 222. The radiation from radioactive elements was found to be of three kinds, called alpha, beta, and gamma rays. Becquerel found that beta rays were negatively charged particles moving at high velocity. Ernest Rutherford identified the alpha rays in 1899 as slower-moving, positively charged particles. The gamma rays, as was the case with X rays, were not deflected by a magnetic field, and their nature was still a mystery. Gamma rays were far more penetrating than X rays and could be detected after passing through 30 inches of iron. There was growing suspicion that they were electromagnetic waves, but Bragg believed that they were neutral particles. Only after von Laue in 1912 proved the electromagnetic nature of X rays did it become apparent that gamma rays were also electromagnetic.

SCATTERING OF ELECTRONS

About the year 1903, Lenard had used thin metal foils to allow electrons from a Crookes tube to escape to the outside and had noticed a scattering of the rays as they emerged from the foil. The scattering was different for each of the metals used and the rays formed patterns that varied with the kind of metal and the velocity of the particles. He deduced from these experiments that most of the mass of the atom was concentrated in a small part of the total volume of the atom.

In 1911 Ernest Rutherford, who was at that time professor of physics at Manchester University, announced the results of a long series of experiments on the scattering of alpha rays, later identified as helium nuclei, after passing through metal foils. Rutherford's associates during these experiments, who carried out most of the work, were H. Geiger and E. Marsden. The source of the alpha rays was a bit of radioactive radon enclosed in a block of lead with only a small aperture on one side through which the rays could emerge as a narrow beam. A piece of metal foil was placed in the path of the rays, and behind the foil was a small microscope mounted on an arm whose axis was the center line of the foil. At the end of the microscope was a fluorescent screen that showed scintillations when struck by the alpha particles.

The experimenters counted the number of scintillations per unit of time at various angles from the line of the beam, deriving a pattern in which the dots representing observed scintillations were more or less symmetrically placed

Figure 15.4 Ernest Rutherford (*From Smithsonian Institution*)

around the center. Most of the rays were not deflected at all or very slightly. Rutherford assumed that the widely deflected rays had collided with one or more atoms.

By means of very complicated reasoning and calculations Rutherford concluded that an atom is largely empty space and that most of its mass is confined to the positively charged center, later called the nucleus, which he calculated had a radius of less than 10^{-12} centimeter. He concluded that the positive charge on the center was exactly equal to the sum of the negative charges on the associated electrons.

PHOTOELECTRIC EFFECT

The term photoelectric effect is somewhat ambiguous. Usually the expression means the photoemission effect, in which electrons are freed from metallic surfaces by the action of light. There are, however, two other photoelectric effects, called the photovoltaic effect and the photoconductive effect.

The photovoltaic effect was discovered in 1839 by Alexandre E. Becquerel, father of Antoine Henri Becquerel. He had immersed two electrodes of the same material in an electrolyte and found that when light shone upon one of them there was a difference of potential between them of about 0.1 volt.

In 1878 R. E. Day found that selenium exhibited a photovoltaic effect. The photovoltaic properties of cuprous oxide were discovered in 1930 by B. Lange, L. O. Grondahl, and W. Schottky. Photovoltaic light meters using silicon were introduced in 1954 by D. M. Chapin, G. S. Fuller, and G. L. Pearson, following a suggestion by Paul Rappaport.

Willoughby Smith was the first to observe photoconductivity in 1873; he found that small rods of selenium became better conductors of electricity when they were exposed to light. Other materials, such as cadmium sulfide, were found to have similar properties by B. Gudden and R. Pohl in the 1920s.

Heinrich Hertz of Karlsruhe, Germany, observed in 1887 that ultraviolet light falling on the electrodes of a spark gap would cause the spark to jump a greater distance. Hallwachs found, in the following year, that ultraviolet light falling on a negatively charged body would cause that body to lose its charge very rapidly, whereas a positively charged body was not affected. Ten years later J. J. Thomson and P. Lenard, one of Hertz's students, discovered independently that light falling on a metal surface would cause the emission of negatively charged particles, later identified as electrons and called photoelectrons.

For most metals only ultraviolet light will cause photoemission. The exceptions are the alkali metals such as sodium, potassium, lithium, cesium, and rubidium.

PLANCK'S CONSTANT

The same period that produced the important findings of cathode rays, radioactivity, and photoemission also brought forth a new concept relating to the energies of electrons or photoelectrons. Max Planck of Berlin had been studying blackbody radiation during the year 1900, and in the course of these studies he derived a theoretical formula that fitted his graphs of experimental results, giving the distribution of radiated energy over all wavelengths for different temperatures.

In the derivation of this equation he found it necessary to assume that radiant energy was emitted in very small discrete bundles called quanta and that the magnitude of a quantum of energy was $h\nu$, in which ν was the frequency of the radiation and h was a new universal constant that he called the quantum of action. The value of this new constant was determined by Robert A. Millikan to be 6.62×10^{-27} erg second.

PHOTOELECTRONS AND EINSTEIN'S EQUATION

During the period from 1897 to 1905, a number of experimenters made measurements of the velocities of cathode rays, Lenard rays, photoelectrons, beta rays, and rays from heated metals. Foremost among these experimenters was J. J. Thomson. Others were Lenard, Becquerel, Seitz, Stark, and Reiger. In carrying out these experiments, various indirect methods were used including magnetic deflection, electrostatic deflection, heating effects, and acceleration or retardation in electric fields.

Albert Einstein, whose fame rests chiefly on his theory of relativity and his hypothesis regarding the mutual convertibility of matter and energy, made an important contribution to the theory of photoelectrons in the year 1905. His proposition in this regard is expressed by the equation $E = h\nu = W + \frac{1}{2}mv^2$. In this equation the energy E equals the product of the Planck constant h and the frequency of the photon ν (the Greek letter nu), which equals the sum of W, the work required to free the electron from the metal, and its kinetic energy $\frac{1}{2}mv^2$. Here m is the mass of the electron, and v is its speed after being freed by the photon. The value of v is determined by one or more of the methods mentioned in the previous paragraph.

The work W is a variable depending on the kind of metal used. It is least for the alkali metals and greatest for the heavier metals, as might be expected, because of the greater force holding the electrons in the heavier metals. The velocity of the photoelectron increases with the energy of the photon. High-frequency photons have greater energies than low-frequency photons, and in order to free electrons from the heavier metals ultraviolet light is required. Adding to the intensity of the light does not change the velocity of the photoelectrons but increases their number.

HYDROGEN SPECTRA

In 1885 J. J. Balmer studied the visible spectrum of hydrogen and derived the formula $\lambda = 3645.6\ [n^2/(n^2 - 4)]$ in which λ (the Greek letter lambda) is the wavelength of a bright line in angstrom units (one angstrom = 10^{-8} centimeter), and n is a whole number 3, 4, 5, ... It was shown that the empirical formula also applied with considerable accuracy to lines in the ultraviolet region. These lines of the hydrogen spectrum have since become known as the Balmer series. Rydberg modified the formula by expressing it in terms of frequency rather than wavelength giving the following expression: $1/\lambda = 109{,}678\ [(1/2^2) - (1/n^2)]$ in which $109{,}678 = R$, the Rydberg constant.

STRUCTURE OF THE ATOM

John Joseph Thomson (1856-1940), head of the Cavendish Laboratory at Cambridge University (not to be confused with William Thomson, Lord Kelvin), mentor of Ernest Rutherford, had conceived of the atom as a positively charged mass in which the electrons were embedded. This concept is sometimes referred to as the plum-pudding model. Such a model failed to explain many of the known facts so that Rutherford was forced, rather reluctantly, to the conclusion that the negatively charged electrons could be kept away from the positively charged center only by centrifugal force. This concept, however, led to another problem that was not resolved until some years later;

namely, a revolving electron, according to Maxwell's equations, should radiate energy.

At this point Niels Bohr (1885–1962), a native of Copenhagen, studied for one year at Cambridge under J. J. Thomson and the following year (1912) at Manchester under Rutherford. He returned to Copenhagen in 1913, where he completed and published his epochal theory concerning the structure of the hydrogen atom. Bohr's model of the hydrogen atom was based on Rutherford's proposal that the electron revolves about a positively charged nucleus. He assumed that in any atom there were as many orbiting electrons as there were positive charges on the nucleus, and in order to overcome Rutherford's objection, he postulated that as long as an electron remains in a given orbit there is no radiation of energy. He theorized further that an electron may jump from one orbit to another. If it moves to the next higher orbit, one quantum of energy is absorbed, and if it jumps to the next lower orbit, one quantum of energy is radiated.

Bohr showed that in the hydrogen atom, which had only one positive charge and one electron, the electron had five or more possible orbits in which the forces within the atom were in momentary equilibrium. The radii of these orbits were related to one another in the ratio of their squares or as 1, 4, 9, 16, and 25. Normally the electron in the hydrogen atom is in the lowest orbit, or ground state. When it is in a higher orbit, the atom is said to be excited. If sufficient energy is supplied to the atom from the outside, electrons are not only sent into higher orbits but may be freed from their atoms altogether as is the case in photoemission.

By inserting the value of h in his equations Bohr determined that the innermost orbit of the hydrogen electron had a radius of 0.5×10^{-8} centimeter and the higher orbits had dimensions in proportion to the squares of their orbit numbers. Furthermore Bohr was able to determine mathematically the wavelengths of the emission lines in the hydrogen spectrum. His calculations agreed with the experimental results obtained by Balmer, and in addition Bohr calculated the wavelengths of the Lyman series of spectral lines in the ultraviolet and the Paschen series in the infrared, both of which had been determined experimentally in 1908. The Brackett and Pfund series in the infrared were found later. In all cases Bohr's calculations agreed with the experimental results and established the truth of the Bohr theory almost beyond doubt.

HEAVIER ATOMS, ELLIPTICAL ORBITS, AND SPIN

For atoms more complex than hydrogen, theoretical solutions for spectral lines are all but impossible, because of the interactions of the orbiting electrons and such further complications as elliptical orbits and spin. In the heavier atoms the radii of the orbits are smaller because of the greater positive charge on the nucleus. Despite these difficulties there could be little doubt as to the validity of the Bohr theory and its applicability when it was extended to include the heavier atoms.

Close examination of the spectral lines of hydrogen and other elements, particularly those of the alkali metals, revealed that many of the lines were multiples, so close together that it was difficult to separate them. Sommerfeld advanced the theory that in some instances these multiple lines were due to elliptical orbits of the electrons. In 1925 Goudsmit and Uhlenbeck proposed the idea that some of the lines, which were doublets, could be accounted for by assuming that some of the electrons were spinning about their axes.

THEORETICAL AND EXPERIMENTAL PHYSICS OF THE 1920s

The period of the 1920s was one of tremendous progress in the development of atomic physics and quantum mechanics. In 1921 A. H. Compton discovered what is called the Compton effect. He found that when X rays were allowed to strike a carbon block other X rays of lower frequency appeared. These new X rays had different frequencies in different directions. Their lower frequency resulted from the loss of energy caused by each collision between an X ray, which behaved much like a particle, and an electron of a carbon atom.

Louis de Broglie of France in 1922, advanced the theory that atomic particles have wave properties. C. J. Davisson and L. H. Germer expanded de Broglie's theory in 1927.

The year 1925 was filled with announcements of new theories in the field of atomic physics and quantum mechanics. W. Pauli set forth a rule in that year called the Pauli exclusion principle which stated that in any atom no two electrons may have the same quantum numbers. E. Schrödinger of Germany developed the theory that an orbiting electron is a wave packet made up of standing waves pulsing within the atom. The Schrödinger picture of the atom was extended by W. Heisenberg in 1925, by Born in 1926, and by P. A. Dirac in 1928. The theories developed by this group were expressed in mathematical terms and were found to be consistent with the Bohr theory. More than forty years have passed since these theories were announced, and they are still generally accepted as valid.

Bohr's original concept of the atom was rather simple, but with the passage of time the picture has become more and more complicated. In 1912, about a year before the publication of Bohr's theory, H. G. Moseley, a young Oxford graduate, using the X-ray spectrograph, found a relationship between the X-ray spectra of various elements and their atomic weights. On the basis of this discovery he deduced that the electrical charges on the nuclei of atoms were of the same order as their atomic weights. It had long been suspected that most of the mass of an atom was concentrated in its nucleus. No longer was the nucleus of the atom thought to be made up entirely of positively charged protons, but it was believed to contain neutrons as well. This new concept of the atom was soon to become very important.

OTHER SUBATOMIC PARTICLES

Theories as to the nature of matter were changing rapidly, especially after the finding of new particles. In 1932 J. Chadwick of England discovered the neutron, and in the same year C. D. Anderson discovered the positron. The neutron was described as a particle similar to the proton but without a charge. Chadwick's discovery was aided by earlier research by Bothe and Becker of Germany and Irène Curie and her husband, Frédéric Joliot. The positron was a very small particle, similar to the electron but with a positive charge. Soon after the discovery of the positron, J. R. Oppenheimer succeeded in producing an electron pair, positive and negative electrons, by the use of powerful gamma rays. A great number of other subatomic particles have been postulated to explain various phenomena, some of which perhaps exist and others of which may be different manifestations of more familiar particles.

Awareness of the existence of cosmic rays started around the beginning of the century. Their source is in outer space, but their exact origin is unknown. They consist of protons, alpha particles, and a smaller number of the nuclei of heavier atoms. They have exceedingly high energies much of which is lost as the rays travel through the atmosphere. As they collide with air molecules thousands of secondary cosmic rays are produced. In the course of this bombardment, positive and negative electron pairs are produced that reunite immediately and are annihilated. A very interesting particle that is a product of cosmic ray collisions is the pi meson. The pi meson may be positively or negatively charged or neutral. It has a mass 275 times that of an electron, and it may achieve a velocity close to that of light. Charged pi mesons disintegrate in a small fraction of a second to form charged mu mesons and a particle called a neutrino. The mu meson, in turn, decays in another small fraction of a second, producing an electron and two neutrinos. The neutral mesons are very unstable and disintegrate even more quickly than the charged mesons into two gamma rays.

Since this is a history of electricity and magnetism, it would not be appropriate to pursue the subjects of atomic physics and quantum mechanics in great detail. Although these studies had their origins in electrical phenomena, they have now extended into regions only remotely connected with the purposes of this book. The remaining discussion of these matters will therefore be brief and will be in the nature of an outline.

THE ELECTRON MICROSCOPE

De Broglie's and Schrödinger's hypotheses of the early and middle 1920s indicated that electrons behave in some respects like particles and in other ways like waves. Hans Busch of Germany showed in 1926 that an electron beam could be focused by an electrostatic or magnetic field. Ernst Bruche and H. Johannson, also of Germany, in 1932 succeeded in producing an image of a heated cathode, using electrostatic lenses. Two years later, Max Knoll and Ernst Ruska of Germany constructed the first electron microscope equipped with magnetic lenses. Siemens and Halske built an electron microscope in 1935 based on a design by von Borries and Ruska, which had a resolving power of 100 angstroms.

In North America, work on the electron microscope was begun in 1935 at the University of California and at Bell Laboratories. RCA Laboratories began its work in 1936 under the direction of Dr. V. K. Zworykin. In the same year work was also started at the University of Toronto.

Electron microscopes may use either magnetic or electrostatic lenses, but very few have been built with electrostatic lenses. Electron microscopes have been built commercially by Siemens and Halske and several other European manufacturers. In the United States they have been produced by RCA and General Electric.

In an electron microscope the entire system is enclosed in a chamber that has been exhausted to a pressure of 10^{-7} atmosphere. Since the specimen to be observed is also enclosed in the vacuum chamber, its moisture must be removed. Because of this, biological specimens are greatly altered, and the usefulness of the instrument is thereby diminished.

The electron microscope is one of the descendants of the Crookes tube. The source of the electrons in the microscope is a hot cathode that furnishes many more electrons than the cold cathode of the Crookes tube. In order to propel the electrons toward the lower end of the tube, it is necessary to provide a potential difference of from 10 to 100 kilovolts between the cathode and the target. The electron beam is brought to a focus at the specimen by means of a toroidal magnet that behaves somewhat like a lens. The beam, after passing through the specimen, is brought to a focus at the screen, which is a plate coated with a fluorescent material. A photographic film may be substituted for the screen. An ordinary optical microscope is capable of magnification up to 2000 diameters, while an electron microscope may magnify up to 200,000 diameters, with a resolving power of about 10 angstroms.

RADIATION DETECTORS

Research in the field of radiation of both particles and electromagnetic waves was largely dependent upon the invention of suitable devices for the detection of such radiation and the showing or measurement of its effects. There is now a long list of such devices including the electroscope, the photographic film, the spinthariscope, the ionization chamber, the Geiger-Müller counter, the cloud chamber, the fluorescent screen, the scintillator, the bubble chamber, radio wave detectors, the radiometer, the thermometer, and the phototube. All of these and others have played an important role in the amazing list of discoveries which followed the invention of the Crookes tube and the discovery of radio waves, the X ray, and radioactivity.

Radio wave detectors were mentioned in an earlier chapter, and others such as the electroscope, the radiometer and the thermometer have been used for many years. One of the most useful detectors for radioactivity has been the cloud chamber, invented by C. T. R. Wilson of Cambridge University in 1912 or perhaps somewhat earlier. In the cloud chamber, the tracks of moving charged particles become visible as they cause the formation of tiny droplets of water in a cloud of water vapor in a glass vessel. A cloud is formed in the vessel above the water in the bottom when the pressure is suddenly decreased by means of a bulb attached to the vessel.

An ionization chamber is a small box or glass tube into which electrified particles are introduced, causing the enclosed gas to become ionized. When a battery and galvanometer are connected between the two electrodes, a weak current will flow through the ionized gas. This current will be proportional to the amount of ionization and the battery voltage and will indicate the amount of ionization.

The Geiger counter was originally invented by Hans Geiger between 1908 and 1913 and later was much improved by Geiger and H. Müller. The Geiger-Müller counter consists of a partially evacuated glass tube containing a short copper tube through the center of which a wire is stretched. The positive terminal of a high-voltage battery is connected to the copper tube, and the circuit is completed through a galvanometer and the central wire. In more modern instruments, the galvanometer is replaced by an amplifier and telephone receiver in which every ray that reaches the counter is registered by a click.

The bubble chamber is a much later modification of the cloud chamber, in which a liquid heated to near its boiling point is used in place of saturated water vapor. An electrified particle passing through the liquid leaves a trail of bubbles.

One of the better and more recent detectors is the scintillation counter; it is a combination of a screen that will emit flashes of light when struck by alpha, beta, or gamma rays or by all three and a photomultiplier tube. The screen is placed in close proximity to the cathode of a photomultiplier tube that is highly sensitive to light, especially in the blue or violet region. A tiny flash on the screen will liberate one or more electrons from the cathode, depending on the intensity of the flash. Below the cathode is a series of screens arranged in zigzag fashion, positively charged, and coated with a material that will give up electrons freely when struck by other electrons. As its name indicates, the photomultiplier tube produces many more electrons than the original number, so that the collector at the bottom of the tube receives an avalanche of electrons. A scintillation counter is a very sensitive detector that is also very fast, with a resolving time ranging from 10^{-5} to 10^{-9} seconds.

ACCELERATORS AND ATOMIC RESEARCH

Prior to 1932, the only method available for the disruption of atoms was bombardment with alpha rays from radioactive materials. Physicists generally were convinced that far better results could be obtained by accelerating atomic particles by artificial means. The energy of alpha rays from radioactive substances was limited, and the particles were projected in random directions, so that they could not be concentrated on a desired target.

Rutherford, who had observed nuclear disruption with the use of alpha rays in 1919, laid plans in 1930 for the construction of a million-volt generator, to be built at the Cavendish Laboratory for use with an accelerator. Two of Rutherford's assistants, Cockroft and Walton, became impatient and in 1932 completed a smaller accelerator, using 150,000 volts. With it they succeeded in projecting protons from hydrogen at a velocity sufficient to disrupt the nucleus of a lithium atom, an event which was accompanied by the release of energy.

Similar activities were under way at other universities. At Princeton, R. J. Van de Graaff in 1931 completed a belt-type electrostatic generator that produced a potential of 1.5 million volts. Van de Graaff transferred to M.I.T. the following year, where he collaborated with K. T. Compton in the construction of a larger machine. They completed a 5.1-MV generator in 1936 that was immediately put to work on nuclear research. In order to overcome corona and moisture difficulties, more powerful generators were later housed in pressurized steel buildings. The highest voltage achieved in a Van de Graaff generator was 12 MV.

At the University of California, at the same time as these other projects, work was in progress on a different type of accelerator, called the cyclotron. This project began in 1930 under the direction of E. O. Lawrence. The cyclotron consisted of a flat, circular vacuum chamber inside of which were two hollow semicircular electrodes, called dees. The vacuum chamber was mounted between the poles of a powerful electromagnet. Inside the chamber, near the center, was a source of hydrogen ions that were accelerated by the high-frequency electrostatic field from the dees. The magnetic field caused the ions to travel in a spiral of increasing radius at enormous velocities until they were finally ejected through an aluminum window. The first cyclotron was completed in 1932 and yielded only 0.8-MeV ions, but a 60-inch machine, completed in 1939, produced 40-MeV ions. The abbreviation MeV stands for million electron volts. One electron volt is the energy acquired by one electron when accelerated through a potential difference of 1 volt.

At the University of Illinois, D. W. Kerst conceived and built an accelerator called the betatron because it produced a powerful beam of electrons, or beta rays, whereas the cyclotron yielded a beam comparable to alpha rays. Several previous attempts had been made to construct an electron accelerator, the earliest of which was made by J. Slepian in 1922, followed by Breit and Tuve in 1927, and Wideröe in

1928. Still other attempts were made by Walton, Jasinsky, and Steenbeck. The Illinois betatron was completed in 1940 and produced 2.3-MeV electrons.

The betatron consists of a doughnut-shaped vacuum chamber made of glass or porcelain, containing an electron gun, all of which is mounted between the poles of a strong alternating magnet operating at 60 to 180 cycles per second. Short bursts from the electron gun are timed to correspond with the proper phase of the magnetic cycle. All of the electrons travel in a single path at the center of the tube, and herein lay the difficulty experienced by the earlier experimenters, who were unable to keep the electron stream stable. The electrons may spin around through the tube as many as 200,000 times in a quarter cycle of the magnetic field and travel at velocities approaching that of light. Not only is the betatron a source of high-speed electrons, it is also used to produce hard and very penetrating X rays. The largest betatron was built by the General Electric Company about the year 1946. It had a magnet weighing 400 tons and produced electrons with energies of 340 MeV.

Another accelerator called the synchrotron is a cross between a cyclotron and a betatron. It has a doughnut-shaped evacuated tube like the betatron, but inside the tube, in addition to the electron gun, are tubular electrodes corresponding to the dees of the cyclotron. Acceleration of the electron is begun on the betatron principle using the alternating electromagnet after which further acceleration is produced by the electrostatic field from the electrodes on the cyclotron principle. Some of these machines accelerated electrons to 99.99 percent of the speed of light with an increase in mass of 600 times the rest mass.

A modification of the synchrotron, built at the CERN laboratory in Geneva in 1959, was used for accelerating protons and was capable of energies up to 28 BeV (billion electron volts). Another such machine, called the cosmotron, was put into service at the Brookhaven National Laboratory in 1960 and was capable of producing 33 BeV.

The fifth important type of accelerator is called the linear accelerator because, as the name implies, the accelerated particles move in a straight line. Ising of Sweden proposed such a device in 1925 but did not actually build one. E. Wideröe, of Brown Boveri in Switzerland, built a small linear accelerator with three electrodes in 1928. It was followed a short time later by a machine with ten electrodes built by D. H. Sloan of the University of California. Others who worked on the problem were J. W. Beams of the University of Virginia, who built a linear accelerator yielding 1.3 MeV. W. W. Hansen of Stanford University began his work on linear accelerators in 1930 and as a by-product invented the rumbatron, from which Russell and Sigurd Varian developed the klystron oscillator tube in 1937. After World War II, L. Alvarez and W. K. H. Panofsky of the University of California attacked the problem, and work was also resumed at Stanford University. On September 9, 1967, a 10,000-foot linear accelerator became operational at Stanford University. It was built by the university under the sponsorship of the U. S. Atomic Energy Commission.

16

Microwaves, Radar Radio Relay, Coaxial Cable, Computers

MICROWAVES

Microwaves are radio waves with wavelengths lying in the electromagnetic wave spectrum between 1 meter and $^1/_{10}$ centimeter. They have properties quite unlike the radio waves used for broadcasting. These short waves are much more directional, are propagated in straight lines like light, and can be confined within a tube or waveguide. Because of their very short wavelengths, microwaves require special equipment for their generation and reception. Ordinary radio circuits have components of capacity and inductance that render them entirely unfit for microwaves.

Microwaves have been put to various uses, the most important of which are radar and radio relay telephone circuits.

RADAR

Radar was very largely a development resulting from defense requirements in World War II, although early forms of radar were in existence as early as 1934 and were in use along the British east coast several years later.

The conception of radar was not the invention of any single individual or group but rather came into existence by steps with contributions by many research scientists. The pioneers in microwave development were Dr. Irving Wolff of RCA and I. E. Mouromtseff of the Westinghouse Company in the early 1930s.

An essential part of pulse radar, the most common form, was the cathode-ray tube, whose ancestor was the Crookes tube. As was mentioned previously in connection with television, the cathode-ray tube was the invention of Karl F. Braun, who applied it to his oscilloscope in 1897, and it was this tube that became the receiving screen of radar and of television.

The word radar was coined by S. M. Tucker of the United States Navy from the name Radio Detection and Ranging. Radar is essentially a method of detecting the presence of, and the direction and distance of, objects not ordinarily visible because of darkness, fog, or distance. It also has the capability, in another form, of determining the speed of moving objects and the direction in which they are moving. These results are accomplished by reflecting microwaves from the objects under observation and receiving the reflected signal by suitable means. In wartime use, radar served not only to detect enemy ships and planes but was used also to direct gunfire with incredible accuracy.

The first discovery leading to the development of radar was made in 1922 by engineers of the U. S. Naval Research Laboratory when they observed that radio waves were reflected back to the transmitter by a ship passing between a sending and receiving station. Beginning with that observation and continuing until 1930, the laboratory performed many experiments on the same phenomenon and constructed primitive forms of radar using ordinary radio waves.

EARLY BRITISH DEVELOPMENTS AND INSTALLATIONS

During the 1930s when Britain was already aware that it might be vulnerable to attack from the continent, early forms of radar were installed along the eastern and southern coasts of the island. In 1936 there were five such stations;

in 1937 there were twenty, and more were added in 1938 and 1939. These installations used a wavelength of 1.5 meters, the shortest length which was possible by the then-existing equipment. The person chiefly responsible for this work was Robert Alexander Watson-Watt. The British realized that to be most effective radar required much shorter wavelengths, especially for airborne radar. The problem was submitted to a research team at the University of Birmingham, which about the year 1940 produced an early form of the magnetron oscillator.

AMERICAN WARTIME RESEARCH AND DEVELOPMENT

When the war began in 1939, various of the research laboratories of American companies took up the radar problem in earnest. Among those working on this problem were RCA, General Electric, Westinghouse, Bell Laboratories, Western Electric, Philco, Raytheon, Hazeltine, Sperry, Federal Telephone, Bendix, and others.

In 1940 the Radiation Laboratory was established at M.I.T., which had on its roster most of the men in the United States who had knowledge of microwaves. This organization worked closely with its British counterparts and with the various radio manufacturers to develop and produce as quickly as possible the most effective radar equipment within their capabilities. The British contributed their version of the magnetron tube, which was improved upon and was soon produced in quantity by the General Electric Company. Between the work of the Radiation Laboratory and the research organizations of the manufacturing companies, progress was little short of miraculous, so that what might have taken years to develop was done in months. The elapsed time between laboratory experiment and operation on the war front was sometimes a matter of weeks.

NEW OSCILLATORS AND OTHER TUBES

Radar made necessary the production of hundreds of new types of vacuum tubes unlike anything previously manufactured. The reason was, of course, that the inherent capacitance and inductance of the older tubes rendered them unfit for microwaves. Furthermore, the transit time of electrons moving from cathode to plate in an ordinary tube was so great that the grid voltage might be reversed before the electron had traversed more than a part of the required distance, and the tube would become inoperative. The parts of the microwave tubes had to be made much smaller and closer together than in the standard tubes.

There was an even greater problem involved in the manufacturing of oscillator tubes because these were required to put out rather large amounts of power and still have internal characteristics suitable for microwave operation. Three general types of such tubes were made: the lighthouse tube, so-named because of its shape, for smaller power output; the magnetron for large power output; and the klystron, a versatile kind of tube with a wide frequency range, used for both transmitters and receivers.

The lighthouse tube can operate up to 4000 megacycles, at relatively low power. The magnetron, which uses a strong permanent magnet to cause the electrons issuing from the cathode to move in circular orbits within several chambers, is capable of producing waves with wavelengths of 1 centimeter or less and can reach peak outputs measured in megawatts. The klystron like the magnetron has cavity resonators but is adjustable over a much wider range and is therefore suitable as an oscillator tube for FM as well as microwave transmission and reception.

TYPES OF RADAR

In the earliest radar sets the pulse system was used, and this system is still the most common. Signals from the transmitter are beamed toward the target in pulses only a few microseconds in length, each of which is followed by a pause of much greater duration. The pulse repetition frequency (PRF) may vary from 60 to 4000 per second. The pulse repetition time (PRT) may therefore vary between 16,667 and 250 microseconds (millionths of a second). The duration of a single pulse is often only a fraction of 1 percent of the time between pulses, which means that the peak output of an oscillator tube may be several thousand times its average power.

In most cases radar is used to scan the entire 360 degrees around the transmitter, and for that reason a rotating or sweep antenna is used. The return signal is generally received by the same antenna, which in the meantime has rotated through a small angle depending upon the distance of the target. The received signal is heterodyned, amplified, and demodulated, after which the pulses are fed into a cathode-ray tube where the target will appear in white against a greenish background. The screen is provided with suitable markings, to indicate distance and orientation.

Two other types of radar have been developed called FM, and frequency shift radar. In FM radar the frequency changes constantly through a predetermined range, corresponding to the pulse in pulse radar. The distance of the target is determined by the beat frequency between the incoming and outgoing signals, and may be read directly on suitably calibrated instruments.

Frequency shift radar is of the pulse type, designed specifically to measure the speed of a moving target by utilizing the Doppler effect. The reflected signal from the moving target is received by the same antenna as the outgoing signal. The received signal is then compared with the constant frequency of an oscillator other than the transmitter, and by means of suitably calibrated instruments the speed of the target can be read directly. This is the type of radar used by highway police.

OTHER USES OF RADAR

The use of radar during World War II helped save England and contributed to the destruction of the enemy. Since that time the military uses of radar have been greatly expanded and refined. Many peacetime applications have been introduced, such as marine navigation, airport landing devices, police radar, and weather forecasting.

TELEPHONE RADIO RELAY

Perhaps the most important peacetime application of microwaves was in the telephone industry for what is called Telephone Radio Relay. Ordinary long radio waves had been used for a number of years for voice communication, facsimile transmission, teletypewriter service, network television and radio, and telemetering.

Telephone Radio Relay was first introduced in the United States by the Bell System in 1947, using the 4000-megacycle band. In such a system microwaves are sent from a terminal, through a directional antenna, to a tower some miles distant. Here the signal is amplified and may be changed in frequency, after which it is sent on to a second tower, and so on, until it is received at the distant terminal. The relay towers are about 30 miles apart and are purposely set in zigzag fashion to prevent what is called overreach, which means the picking up of a signal by a more distant tower instead of the nearer one for which it was intended.

FREQUENCY BAND ALLOCATIONS

In 1949 the Federal Communications Commission allocated three frequency bands for telephone radio relay communication: 4000, 6000, and 11,000 megacycles, each of which was 500 to 600 Mc in width. In the early experimental use of the 4000-Mc band there were six channels each of which was capable of providing 480 circuits. In 1959 the same frequency band was divided into twelve channels, and more recently each of these channels was made capable of providing 900 circuits, making a total of 10,800 circuits in the 4000-Mc band alone. In practice two of the channels are reserved for standby, leaving 9000 working circuits. Still more recent improvements, involving the use of transistors, will increase the capacity of each channel to 1200 circuits.

When the 4000-Mc band began to approach the limit of its capability, work was begun on the use of the 6000-Mc band, which was divided into eight channels, two of which were standby. Each channel furnished 1800 circuits, making a total of 10,800 working circuits. Both the 4000-Mc and 6000-Mc bands have been used principally for long-distance service, between important terminals, but more recently a portion of the 6000-Mc band has been used for what is called short-haul service.

The capabilities of the 11,000-Mc band have probably not been fully developed up to this time. This band is less efficient than either the 4000- or 6000-Mc band for the reason that the very short waves have a shorter range and are weakened by the presence of raindrops in the atmosphere. The 11,000-Mc band has therefore been used primarily for short-haul service, especially in dry areas. It can supply 600 circuits over distances of 200 miles in dry regions and 100 miles elsewhere, but repeater stations are much closer than with the longer wavelengths.

To provide the large number of circuits per channel the microwave is modulated by the signal and by a carrier wave, and in addition the outgoing signals are polarized horizontally and vertically. At the receiving end the signals are filtered out and reappear as ordinary telephone messages in the telephone exchange. The relay towers have two sets of antennas, one for sending and one for receiving, and a two-way telephone conversation follows two different paths between stations.

COAXIAL CABLE

Coaxial cable was invented by two Bell System engineers in the 1920s: Herman A. Affel and Lloyd Espenschied, who applied for a patent in 1929. The capabilities of such a cable were demonstrated experimentally in 1936 by a line between New York and Philadelphia. In 1941 the first commercial coaxial cable was laid between Minneapolis and Stevens Point, Wisconsin, a distance of 195 miles. This installation originally provided 480 two-way telephone circuits. This cable contained four coaxial tubes; two of which were for standby.

A coaxial tube consists of a 3/8 -inch copper pipe in the center of which is a No. 10 copper wire. The wire is held in place by perforated plastic disks spaced about 1 inch apart. At least two coaxial cables are required for a line: one for outgoing and the other for return signals. At the present time cables containing up to twenty tubes are being manufactured. The sheath for such a cable is made up of layers of polyethylene and lead and is sometimes protected by an outer armor of steel tape. Inside the cable and alongside the coaxial tubes insulated wires are placed for use by maintenance crews to enable them to communicate with stations along the route. Generally the cables are buried in trenches, but sometimes they are carried overhead for short distances. The trench also contains ground wires placed on either side of and above the cable for lightning protection.

The original Minneapolis-Stevens Point line operated in the 68- to 2788-kilocycle range with repeater stations 7.8 miles apart. This line had a capacity of 480 voice circuits, later increased to 600. In 1953 a cable was completed between New York and Philadelphia operating in the 312- to 8224-kilocycle range that had a capacity of 1860 voice circuits for each pair. In this installation repeater stations were 4 miles apart. A more recent development was a 20-tube cable, a test installation of which was made between Dayton and Rudolph, Ohio, in 1965. A 20-tube cable was put into service between Washington and Miami in 1967. It has a total capability of 32,400 circuits and operates in the 564- to 17,548-kilocycle range, with repeater stations 2 miles apart.

Experimental work is under way for the use of what is called pulse code modulation for use on coaxial cables, which is expected to increase the capability of these cables still further and will eliminate distortion.

There are at the present time about 15,000 miles of coaxial cables in the United States. Although radio relay handles almost all intercity television program transmission and about two-thirds of other long-distance communication, coaxial cable is growing more rapidly and presumably will in the future account for a larger proportion of telephone traffic. With the introduction of larger coaxial cables the cost per circuit has been decreasing until it is now about

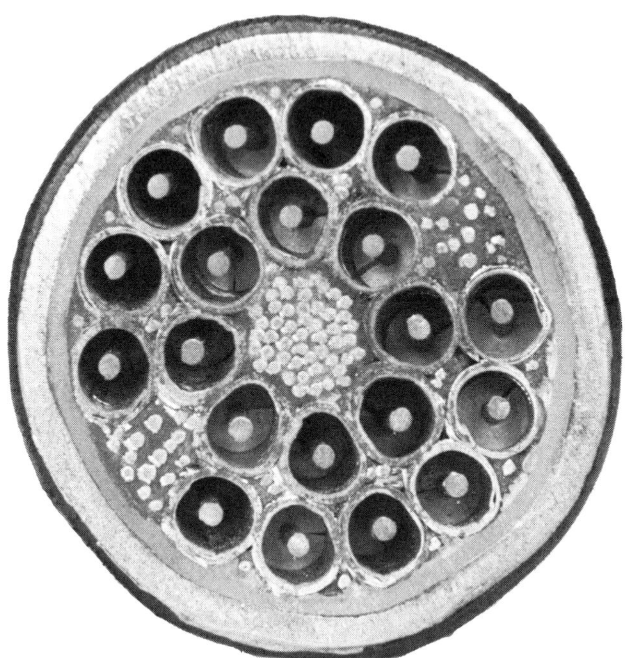

Figure 16.1 Coax-20 Coaxial Cable. This cable, approximately three inches in diameter, handles 32,400 voice channels simultaneously. *(Courtesy Bell Labs)*

the same as radio relay. In general coaxial cable is somewhat more reliable than radio relay but is subject to damage because of being accidentally dug up or plowed up.

The Bell System is also conducting research on the use of laser beams for communication.

COMPUTERS

It is difficult to trace the ancestry of modern electronic computers except to say that they were preceded by various kinds of mechanical calculating devices going back to the abacus, which is still widely used in many countries in the Orient. The earliest form of desk calculator was introduced by Pascal in 1642 and was used only for addition. A modification of the Pascal machine was made by Leibniz in 1671, and this machine was capable of performing multiplication. Charles Xavier Thomas of Colmar in 1820 was the first to make calculating machines commercially. The Thomas machine was not altogether reliable and therefore was not a success. The Brunsviga calculator, invented by Bohdner in 1876, was produced commercially several years later and was very successful. This machine, manufactured in Germany, could add, subtract, multiply, and divide and was the prototype of the many desk calculators of later years.

The slide rule had its beginning with Gunter's logarithmic scale in 1620. The sliding scale was introduced by Wingate of England in 1672. Lieutenant Amédée Mannheim of the French army introduced the modern form of slide rule in 1850.

Probably the earliest integrator device was Amsler's planimeter, which dates back to about the year 1854. It had been preceded by somewhat similar contrivances made by Conradi and Fleischauer.

Computer development was aided by the invention of tabulating machines, using punched cards. The first such machine was invented by Dr. Herman Hollerith of the United States Census Bureau in the middle 1880s. The 1880 census reports were delayed several years because of the vast amount of clerical work involved. Dr. Hollerith foresaw the possibility that future reports, because of their increasing complexity, could not be completed before the beginning of the next census.

He conceived the idea of using punched cards to expedite tabulating the information gathered in the census. The idea was borrowed from the Jacquard loom, in which long sheets of punched cardboard were used to instruct the loom in the weaving of intricate patterns. Even the use of punched cards for statistical purposes was not new, because Charles Babbage, of England, had made such a suggestion as early as 1840 but had not carried out the idea.

Hollerith began work on the punched card system in the middle 1880s and within a few years had produced the equipment that was used successfully in tabulating the 1890 census returns. Business concerns began to adopt the system, and to meet their demands Dr. Hollerith formed a company in the 1880s called EAM for Electric Accounting Machines. The company was reorganized in 1911 as the Computing Tabulating Recording Company, and in 1924 it became International Business Machines, or IBM.

Another company that played an important part in computer development was Remington-Rand Inc., which was formed by the merger in 1927 of the Remington Typewriter Company and the Rand Kardex Corporation. Remington-Rand Inc. acquired the Eckert-Mauchly Computer Corporation in 1949. In 1955 Remington-Rand and the Sperry Corporation merged to form the Sperry-Rand Corporation. Among the many products of the new company are UNIVAC computers, punched card systems, tape, and other computer supplies.

Beginning in the early 1950s many other American firms entered the computer field, including General Electric, Control Data Corporation, Honeywell, Digital Equipment Corporation, National Cash Register, Philco, RCA, Raytheon, and Burroughs. In addition, there are hundreds of electronic companies producing magnetic tape, disks, drums, cores, paper tape, cards, transistors, and the many other items which are used in connection with computers. In addition to the American companies, there are now many companies abroad that are in the computer business.

COMPUTER DEVELOPMENT

Charles Babbage, who was born at Totnes in Devonshire, on December 26, 1792, was the son of a banker. He entered Cambridge University in 1810, where he became an outstanding student of mathematics. He was distressed by the fact that many of the mathematical tables of the time were full of errors. In 1812, with an open book of logarithm tables before him, he conceived the idea of building a computing machine that could be used to produce such tables and that would be free of human error. For some years he spent much of his time in an effort to design such a machine. As a result of his labors he produced a machine in 1822, which he called a Difference Engine, so named because it was designed to calculate such quantities as logarithms and trigonometric functions for use in tables, based upon differences. This machine was capable of calculating the values of simple equations to six decimal places. Babbage's next effort was to build a larger Difference Engine capable of making calculations of such quantities up to twenty decimal places, but he found that machine shop tools and technology of the time were not sufficiently advanced to build a machine of this degree of complexity. In 1833 Babbage designed a different type of machine, which he called his Analytical Engine. This machine, to which Babbage devoted the remainder of his life but which he never succeeded in building, was a mechanical computer. It had four basic parts comparable with those of a modern computer. One of these he called the store, which was a limited mechanical memory. Another part he designated as the mill in which the calculations were carried out. A third part was designed to transfer information between the store and the mill, and the fourth part was the output portion of the machine. Babbage's design called for having the machine print the results.

In 1876 Lord Kelvin described an analog type of computer that was designed to solve differential equations, but this too was never built. The first computer to be completed and put to use, excluding Babbage's Difference Engine, was built by Dr. Vannevar Bush and his associates at the Massachusetts Institute of Technology. This machine, begun in the middle 1920s, was completed in 1930. It was entirely mechanical, cumbersome, and slow, but it was capable of solving complex problems. Its greatest fault was that the time required to set up the machine, or in more modern terms, to program it, was very long. Copies of this machine were built in the 1930s: one for the Aberdeen Proving Ground and another for the Moore School of Electrical Engineering at the University of Pennsylvania.

During the 1930s Dr. George R. Stibitz of Bell Telephone Laboratories designed and built a computer for use in the calculation of alternating-current problems involving the use of complex numbers. This machine used telephone relays instead of gears, cams, and levers employed by the earlier computers. Input to the machine was by means of a keyboard, and the output was delivered by a teletypewriter.

Dr. Vannevar Bush and his co-workers completed a greatly improved computer in 1942 in which many of the operations were carried out by means of electrical relays and switches but which still contained mostly mechanical counting devices.

Dr. Howard Aiken, professor of mathematics at Harvard University, began the construction of a computer in 1937, with the assistance of J. W. Bryce, C. D. Lake, B. M.

Durfee, and F. E. Hamilton of IBM. The project was financed by the U. S. Navy. This machine, called the Automatic Sequence Controlled Calculator, or Mark I, was completed in 1944. It contained electrical relays and switches but was still mostly mechanical. Input and control of the machine were accomplished by the use of punched paper tape, and the output was recorded by means of automatic typewriters. The Mark I computer operated successfully over a period of about fifteen years before it was retired, but compared with later machines it was slow. The calculation of a logarithm, for example, required 90 seconds.

A second computer, called Mark II, was begun by this group in 1945 and completed in 1947. It contained some 13,000 electrical relays and could store 100 numbers of 10 digits each in addition to a sign for each number. Input to the machine was by means of punched paper tape, and paper tape was also used as supplemental intermediate storage. Output from Mark II was recorded by automatic printers or typewriters. Its speed was much greater than that of Mark I, but it was still very slow in comparison with later machines.

DIGITAL AND ANALOG COMPUTERS

Computers are of two general types, digital and analog, but the distinction is now rapidly disappearing. A digital computer, as the name implies, operates upon discrete numbers and performs arithmetic computations. An analog computer is capable of calculating transcendental functions or of solving differential equations. As its name indicates, an analog computer uses analogs such as voltages, resistances, angles, revolutions, lengths, or other measureable quantities to represent the actual amounts. A digital computer is exact, but an analog computer, because of the nature of the quantities involved and the kinds of measurements made, cannot be absolutely accurate.

ELECTRONIC COMPUTERS

The first electronic computer, called Mark III, or ENIAC for Electronic Numerical Integrator and Calculator, was built for the U. S. Army under the direction of Drs. J. P. Eckert and John Mauchly of the Moore School of Electrical Engineering. This machine, begun in 1942, was completed in 1945. It contained 18,000 vacuum tubes and 6000 switches and required about 150 kilowatts of electrical power for its operation. ENIAC was the first computer to use the binary system for all of its operations except the storage section. The storage or memory section consisted of energized vacuum tubes and was capable of storing 200 ten digit numbers.

Following the ENIAC the next important development was the IBM Selective Sequence Electronic Calculator completed in 1948. This machine also used vacuum tubes and relays and operated on the binary system.

In England, where the earliest computer ideas originated with Babbage and Lord Kelvin, work on computers was going on also but at a somewhat slower pace. The EDSAC was completed at Cambridge University in 1949. This was the first computer to use a storage system other than vacuum tubes or mechanical devices and consequently needed far fewer vacuum tubes and therefore less power than its predecessors. Its memory system consisted of acoustic mercury delay lines. The EDSAC contained only 3000 vacuum tubes.

The National Physical Laboratory in London completed the Automatic Calculating Engine or ACE in 1950. It used mercury delay lines and contained only 1000 vacuum tubes. A similar larger machine was built in the late 1950s.

Back in the United States, the Moore School of Electrical Engineering built the EDVAC for the Ordnance Department of the U. S. Army in the early 1950s. It had mercury delay lines for its memory and contained about 3500 vacuum tubes. The EDVAC was designed to handle 44-digit binary numbers.

Eckert and Mauchly formed a company called the Eckert-Mauchly Computer Corporation, which produced the first UNIVAC in 1951. This machine was equipped to use eight-channel metallic magnetic tape for its input, on which the instructions were encoded by means of a typewriter keyboard. Its storage section had both mercury delay lines and magnetic tape. Output was recorded on magnetic tape from which the results could be reproduced in printed form by a special typewriter.

The first computer to use transistors instead of vacuum tubes was the SEAC (Standards Eastern Automatic Computer) built by the U.S. Bureau of Standards at Washington in 1950. Except for the fact that transistors replaced vacuum tubes, the SEAC followed to a large degree the logical design of EDVAC. The SEAC was constructed in place at the bureau, and was put in service in April 1950.

Transistors had not reached the degree of reliability in 1950 that they attained later, and it was therefore not until 1958 or 1959 that computer manufacturers generally adopted the transistor. After the adoption of transistors, it became possible to reduce the size of computers very greatly, and power requirements were cut to a fraction of those of the earlier computers.

Until the early 1950s nearly all computers had been built for government agencies or for universities. Later in the decade and especially after the advent of transistors, industrial and commercial concerns began to purchase computers in increasing numbers. The early computers were custom-built and were often designed for a specific purpose. With the widespread use of these devices came mass production and the designation of computer types by manufacturer's model numbers.

MEMORY SYSTEMS

Every calculating machine must have some sort of memory. It may consist merely of certain mechanical settings, as in the early machines, or energized vacuum tubes, punched cards, punched tape, electrical or acoustic delay lines, minute electrostatic charges, or, most commonly in modern computers, some form of magnetic memory. Magnetic recording had its beginning in a patent issued to Valdemar Poulsen of Denmark on December 1, 1898, covering a method of recording a telephone message on a steel wire by magnetizing the wire in tiny spots along its length to correspond with the undulations of the telephone current. Poulsen had little commercial success with his invention at the time. Following World War I, wire recording apparatus came on the market, but the quality of reproduction was poor. Magnetic tape began to replace steel wire for recording in the late 1930s. Magnetic tape consists of a plastic base of Mylar or cellulose acetate 0.0007 inch to 0.0015 inch thick. The magnetic coating consists of a finely divided oxide of iron mixed with a plastic binder applied to the base to a thickness of from 0.0004 inch to 0.0007 inch. Magnetic disks or drums are made in a somewhat similar way; a binder mixed with an oxide of iron is applied to the surface of a disk or drum.

In order to encode information magnetically on any one of these devices, a small electromagnet called the writing head is used, which is energized by electric currents under the control of punched cards, punched tape, or by other means. The same or a similar small electromagnet, called the reading head, is used to reproduce the encoded information.

Magnetic cores are different, both in composition and method of application, from the other magnetic memory devices. Magnetic cores were invented by Dr. Jay W. Forrester and were first used around 1950. They are very small beads of ferrite material, strung on lattices of intersecting wires. When currents of sufficient strength are sent through these wires, the selected cores are magnetized in the desired direction. A current less than a certain threshold value will not produce a magnetomotive force sufficient to change the direction of magnetization of a core. In order to "pulse" a core a current of slightly more than one-half the threshold amount is sent through each of two wires that intersect at the desired core. The two electromagnetic fields combine at the one particular core resulting in a change in its magnetization. The other cores along both wires receive less than the threshold current and are not affected. This process is called the "coincidence current selection system" and was invented by Dr. Forrester.

The ferrite material used in the cores is an important discovery of Dutch scientists employed by the Philips Company in Holland during the Second World War. Since that time ferrites have found many important applications, not only as cores, but also in high-frequency transformers. Perhaps the simplest way of describing a ferrite of the kind most useful in computer and other applications is to say that it is a compound in which one of the iron atoms in magnetite (Fe_3O_4) has been replaced by a different metallic atom such as Mn, Cd, Co, Zn, Ni, Cu, or Mg. The resulting compound is one having the composition $MnFe_2O_4$, or $NiFe_2O_4$, or other similar compounds with other metals, or in some cases mixtures of metals. Ferrites are not metallic but resemble ceramic materials in appearance, brittleness, and hardness. They are shaped by molding and grinding.

The characteristics of ferrites that make them desirable for use in memory devices and for other purposes are that they have a nearly rectangular hysteresis loop, have high electrical resistance and therefore low eddy current losses, have high permeability, and have low coercive force. Their response to magnetic impulses may be measured in microseconds.

The various kinds of memory systems have inherent advantages and disadvantages involving such things as cost, space required, and most important what is called access time, or the time required to retrieve a given bit of information. Paper tapes, punched cards, and magnetic tapes are relatively cheap, but access time is rather long, because in the case of tapes the entire tape must be scanned for the desired information. Magnetic disks are also inexpensive but are limited in capacity, and access time is rather long. Magnetic drums have shorter access time but are more expensive, and they are bulky. Magnetic cores are the most expensive but have by far the shortest access time, and a vast amount of information can be stored in limited space.

INPUT AND OUTPUT SYSTEMS

The input to a computer may be for the purpose of giving instructions or programming, or may be designed to give information to the internal storage or memory system. The method used may take many forms such as keyboard, punched cards, punched paper tape, magnetic tape, drums, disks, cores, or any number of other methods. In programming a computer it may be given new information, or it may be called upon to make use of its internal memory system.

The output from a computer may also take many forms such as those mentioned for input, or the results may be printed or shown on a screen. The computer may also be used to perform such tasks as check sorting, check writing, directing the operations of a machine or process, billing, bookkeeping, on-line teller service, and hundreds of other operations under the general heading of automation or cybernetics. Utility meter readers now use special pencils to record meter readings, a process called mark sense. Such records can be read by a computer that then prepares the bills and at the same time makes totals of whatever figures are desired.

NUMERATION

Calculators and the earlier computers used the decimal system. As work progressed on these devices it became apparent that a computer using the binary system was considerably less complicated and less expensive to build than one using the decimal system. Computers employ such things as switches, vacuum tubes, transistors, and many magnetic devices, all of which generally have only two states: on or off, positive or negative, or in general what is called flip-flop. Because these devices, as they are used in computers, have only two positions, they are therefore best adapted to the binary system, which uses only the numbers 1 and 0 in a very large number of combinations to represent any rational number. Each binary 1 or 0 in the memory system or elsewhere is called a bit. The number 9 in binary notation is 1001 and therefore consists of four bits.

Dr. Stibitz employed the biquinary system (a combination

of binary and quinary used to produce the equivalent of decimal numbers) in his 1937 computer. Some of the earlier computers had made limited use of the binary system. The ENIAC, which was the first electronic computer, still used the decimal system in its memory, but practically all later computers were entirely binary.

At the time the ENIAC was under construction, Dr. John von Neumann, who was then at Los Alamos, Dr. Herman H. Goldstine, who was at the Aberdeen Proving Grounds, and Arthur W. Burks of the Philosophy Department of the University of Michigan wrote jointly a series of reports on the subject of computers that became classics in computer history. The first of these appeared on June 28, 1946. This series of reports greatly influenced the course of development of computers, as for example, in the matter of stored programs and the use of the binary system.

17

Plasmas, Masers, Lasers, Fuel Cells, Piezoelectric Crystals, Transistors

The behavioral characteristics of free electrons in the three states of matter have similarities as well as differences. A fourth state in which electrical phenomena may occur must be added, namely, a partial or complete vacuum.

PLASMAS

The term plasma has been applied in recent years to ionized gases. A wide range of conditions is embraced under the general term, ranging from gases in the sun and stars, whose temperatures are measured in millions of degrees, to the frigid outer atmosphere of the earth. Plasmas may exist at high temperatures and pressures or at lower values of both, but most of the familiar phenomena occur in the lower ranges.

There is nothing new about the phenomena themselves since they include such things as the experiments of Hawksbee with partially evacuated electrified glass globes, the glow of mercury vapor in the Torricellian tube, the electric arc, flames, lighting tubes using gases, the aurora, and many other examples. Plasmas are usually electrically neutral in that they contain equal numbers of positive and negative ions.

MASERS AND LASERS

A maser is an amplifier or generator of electromagnetic waves. The name is an acronym derived from the descriptive title Microwave Amplification by Stimulated Emission of Radiation. A laser is a similar device except that ordinary light waves are used instead of microwaves. Masers employ gases as a medium, while lasers may use gases, liquids, or solids. It is characteristic of masers and lasers to emit coherent radiation, that is, radiation that is almost entirely of a single frequency and that is of the same phase. The beam consists of rays that are so nearly parallel that the divergence is only about one second of arc.

The explanation of maser or laser action is to be found in Planck's quantum hypothesis, Bohr's theory of the energy levels of electrons in different orbits, and Einstein's postulation in 1917 that it is possible to cause electrons to jump from one orbit to another and in the process to absorb or give up energy in discrete quanta or photons. As an extension of these theories it occurred to several scientists that it should be possible to excite the atoms of a confined gas by applying outside stimulation in the form of electromagnetic waves. The stimulated atoms would then radiate electromagnetic waves of a frequency characteristic of the gas being used. By reflecting these waves back and forth there would be an increasing excitation of the atoms until the action became strong enough to deliver part of the energy externally. The external energy is supplied by the stimulation, and there is therefore no violation of the law of the conservation of energy.

Among those who had been considering such matters were Nicholaas Bloembergen of Harvard, N. G. Basov and A. M. Prokhorov of Lebedev Institute of Moscow, and C. H. Townes of Columbia University. It was Townes who first succeeded in producing an operational maser in 1953. He and his students were already familiar with the microwave emission characteristics of ammonia gas. By stimulating the atoms of ammonia gas contained in a metal box, whose dimensions were such as to produce resonance at the required microwave frequency, they succeeded in producing a strong microwave beam of a single frequency. The action was similar to that which occurs in cavity resonators such as the magnetron and klystron.

In 1952 while Townes and his associates were at work on the ammonia maser, Basov and Prokhorov pointed out the possibility of constructing a "molecular generator" to a conference on radio spectroscopy. In 1954 they published the theory underlying a device such as the maser built by Townes, and in late 1955 or early 1956 they operated such a device. For their accomplishments in this field the Russians shared a Nobel Prize with Townes in 1964.

Maser or laser action depends upon the ability of the apparatus in use to propel a majority of the electrons, in the atoms of whatever medium is employed, into higher orbits. This process is called population inversion. The electrons in the excited state may have been lifted one step to the next higher orbit or two steps to the second higher orbit. In the first case we have a two-level device and in the second case, three-level. The original ammonia maser was two-level. In 1955 Basov and Prokhorov proposed a three-level gas maser, and the following year Bloembergen announced his ideas on a three-level solid state maser.

Following the original ammonia gas maser, there were several years during which there appeared to be little progress in this field, but there was nevertheless considerable research, and many papers were written on the subject. In 1957 H. E. D. Scovil and his associates developed a paramagnetic maser, and in 1958 Makhov, Kikuchi, Lambe, and Terhune obtained maser action in ruby. The most important paper published during this period appeared in the December 1958 issue of *Physical Review* written by Dr. Arthur Schawlow and his brother-in-law, Dr. Charles H. Townes, entitled "Infrared and Optical Masers." The term optical maser was replaced by laser in 1960. The paper dealt with optical masers in general and also proposed the construction of a maser in which the medium was to be potassium vapor. A maser of this type was constructed but it failed.

At a symposium on quantum electronics held at High View, New York, in September 1959, Schawlow proposed an optical maser in which the medium would be a solid in the form of a rod with polished ends and open sides. He used as an example a ruby rod and discussed its possible use in an optical maser.

The actual construction of the first ruby laser was the work of T. H. Maiman of the Hughes Aircraft Company. It contained a synthetic ruby rod only 4 centimeters in length and ½ centimeter in diameter. It was pumped by a spiral xenon flash lamp that emitted short bursts of light of about $1/1000$ second duration followed by intense flashes of red laser light from the rod with a momentary power output of 10,000 watts. This achievement was announced by Dr. Maiman at a news conference in New York on July 7, 1960.

After the first ruby laser there was feverish activity in many laboratories. More powerful ruby lasers were built within a few months, and the field of experiment widened to include many new types. For solid state lasers such materials as neodymium-doped glass, neodymium-doped calcium, calcium tungstate, and dysprosium-doped calcium fluoride were used successfully in place of ruby. Other types of masers and lasers included semiconductor and single- and two-gas types. Liquids were used also.

GAS LASERS

At the Bell Laboratories three young scientists, Ali Javan, Donald R. Herriott, and William R. Bennett, Jr., were engaged in research on a two-gas laser at the time the ruby laser was announced. The American Telephone and Telegraph Company's immediate interest in laser work was the possibility of using such devices in communications. The theory underlying this research had been conceived by Javan, and Herriott and Bennett developed the design of the apparatus, which consisted of a quartz tube 40 inches long filled with a gas mixture, at low pressure, containing ten parts helium and one part neon. Inside the tube at each end were adjustable mirrors to reflect the light back and forth. Pumping was accomplished by means of three electrodes on the outside of the tube that were energized by a radio-frequency generator. This laser first became operative on December 13, 1960, about six months after the ruby laser announcement.

Other combinations of the noble gases were tried successfully and also carbon monoxide, carbon dioxide, and cesium vapor. The gas lasers operate at low temperatures and are especially suited to communications work. They are generally very inefficient, but for the uses to which they have been put the low efficiency is not very important.

APPLICATIONS

Laser applications have been growing at a tremendous rate, and it would be difficult to foresee what the future may hold. Lasers have been built that have momentary power outputs as high as a billion watts and are capable of welding, piercing, or fusing the hardest materials. They can produce temperatures far exceeding those on the surface of the sun. Lasers are finding uses in surgery, dentistry, and biochemical research. One such use is in welding a detached retina. They are also used for making measurements of extreme accuracy, both microscopic and macroscopic, as for example the measurement of earth movements for predicting earthquakes. The frequency rate of lasers is so precise that they can be used to make time measurements that would be almost free of error over periods of many years.

It is believed that in the field of communications it may be possible to transmit millions of messages simultaneously over a single laser beam, but there are still great difficulties to be overcome. Lasers will very probably be another powerful tool in the physics laboratory as well as in chemical and biochemical laboratories. There can be no doubt that mankind will reap rich rewards from laser research and technology.

ELECTROLYTIC AND ELECTROCHEMICAL PHENOMENA

Electrochemistry had its beginning with the invention of the electric battery by Alessandro Volta. That discovery was followed shortly by the important researches of Berzelius of Sweden and the work of Sir Humphry Davy and of Faraday. Since that time great industries have been built upon these discoveries, and there has been steady progress due to the efforts of thousands of researchers.

There have been no spectactular new developments in this field in recent years but rather a constant refinement of older methods. The most noteworthy recent developments have been the production of better types of primary batteries for longer life and greater reliability and new experiments with the fuel cell.

The principle of the fuel cell is not new, because as early as 1839 William Grove of England made such a cell using oxygen and hydrogen for fuel. What is new is the development of fuel cells into practical forms, suitable for certain limited applications. Up to this time the cost has been too great to compete with other forms of electrical generation.

Fuel cell efficiency is high as compared with that of steam or internal combustion engine plants, running as high as 60 to 70 percent, compared with 40 percent or less for the best modern steam plants.

Fuel cells resemble other primary batteries in some respects but differ from them in other ways. Various materials are used for plates in fuel cells such as platinum or platinum-plated carbon. The plates are immersed in an acid, salt, or alkaline electrolyte. Oxygen and hydrogen are fed into the cell where these gases combine to form water and in the process produce an electric current. The action is the reverse of that which takes place in the electrolysis of water. Other fuels such as propane, natural gas, methyl alcohol, kerosene, and gasoline have also been used but at much higher temperatures. Fuel cells using acid or salt electrolytes operate at ordinary temperatures, but cells using certain alkalies may operate at the melting point of the electrolytes.

Up to this time the fuel cell has had very limited applications, the most noteworthy of which has been its use in space capsules, beginning about 1966. The Allis-Chalmers Company experimented with fuel cells in the 1950s for driving a tractor, but not on a commercial basis. Experimentation in this field is still in progress by many companies in the United States and Europe, but nowhere has the fuel cell reached the production stage. Because of its high efficiency there is hope that it may be possible to produce a low-cost design. An inexpensive fuel cell would have a multitude of applications, not the least of which would be the propulsion of vehicles.

PIEZOELECTRICITY

Piezoelectricity is the name given to a phenomenon in which a crystal becomes electrified by pressure. The prefix piezo is from a Greek verb meaning to press. The effect was discovered at the Sorbonne in Paris in 1880 by Pierre and Jacques Curie. Piezoelectricity is closely related to pyroelectricity, a phenomenon discovered by Aepinus in 1756 and defined as electrification by heat that often occurs simultaneously with piezoelectricity.

For many years piezoelectricity was merely an interesting phenomenon, and no practical applications were made of it until 1916 when Langevin attempted to use the principle as a submarine detection device but with little success. During World War I Walter G. Cady of Wesleyan University in Middletown, Connecticut, discovered that mechanical resonance of vibrating crystals produced weak alternating currents of corresponding frequency. This discovery led shortly to the quartz crystal frequency control devices used by radio stations. Beginning in 1925 the Bell Telephone System used crystals to control frequencies for its multichannel lines.

Piezoelectric crystals were applied to microphones and phonograph pickup cartridges, for which purpose it was found that Rochelle salts were admirably suited. Much of the research on these applications was carried on by the Brush Development Company of Cleveland, beginning in 1930.

Around the year 1940 Arthur von Hippel and his colleagues at Massachusetts Institute of Technology discovered the strong piezoelectric properties of crystals of barium titanate ($BaTiO_3$), lead metaniobate ($PbNb_2O_6$), and lead titanate zirconate ($Pb(TiZr)O_3$). These crystals were soon put to use for many of the more common applications, but for frequency control and for wave filters the preferred material was still quartz.

SOLID STATE DEVICES

Some of the most noteworthy accomplishments in electrical science have come in this field in recent years. The term solid state device is often incorrectly used as a synonym for semiconductor. The correct concept is much broader and includes besides semiconductor devices and oxide rectifiers such things as resistors, transformers, reactors, and such other devices as can control electrical currents without moving parts, heat, light, or vacuum gaps.

SEMICONDUCTORS

The distinction between conductors and insulators has been a long-accepted fact, but the reasons for the differences have not been understood until recently, and even today there is a great deal of mystery surrounding them. Exactly how a current of electricity flows through a wire still presents problems. It seems clear that there is a migration of free electrons from atom to atom, impelled by an electromotive force. A displacement of one electron in an atom by another occurs, but it is also clear that this movement of electrons cannot take place at the speed of light, a fact which leads to the conclusion that the electromagnetic field is the real carrier of electrical energy.

Semiconductors, as the name indicates, may be conductors or nonconductors, depending upon the temperature in intrinsic semiconductors or upon the electrical bias imposed upon the crystals from without in impurity semiconductors. Among the most commonly used semiconductor materials are germanium, silicon, and gallium arsenide. In the pure state these substances are nonconductors, but with the addition of very small amounts of certain impurities they can become semiconductors.

The theory behind this behavior may be stated briefly as follows: The binding force that holds atoms together to form molecules and molecules together to form crystals resides in the outer ring of electrons in the atom, and the number of such charges determines the valence of the atom. Normally, when the valence bonds to other atoms are satisfied, the atom has no loosely held electrons in its outer ring. If, however, an impurity is added that has a different number of electrons in the outer orbit, there will be either an excess or a deficiency of electrons in the valence bonds of the crystal. When there is an excess the semiconductor is said to be *n*-type, and when there is a deficiency the semiconductor is *p*-type. In the region where the latter condition exists there is said to be a hole. A junction between *n*-type and *p*-type semiconductors will pass a current in only one direction. The impurities are different for various kinds of crystals. Those most commonly used are arsenic and aluminum.

TRANSISTORS

Bell Telephone Laboratories had been working on the problem of providing a cheaper and better electronic switching device than the vacuum tube for some years with little success. In 1947 two of the researchers at Bell Laboratories, Walter H. Brattain and John Bardeen, succeeded in producing a low-frequency amplifier using a piece of silicon immersed in a salt solution. A month later they substituted a bit of germanium for the silicon and again succeeded in producing amplification. Research continued at an accelerated pace until in June 1948 the first transistor was made by inserting three pointed wires into a piece of germanium. Two of these wires were part of an external circuit, and the third, when connected to a battery, provided a bias voltage in much the same manner as is used with the grid of a vacuum tube. The bias voltage must be sufficient to overcome the counterpressure, or barrier within the crystal. With the required bias voltage a current can be made to flow through the crystal as through any other conductor.

A third member of the Bell Telephone Laboratories team was William O. Shockley, who had developed the theory underlying transistor behavior. In 1949 he published an article in the *Bell System Technical Journal* entitled "The Theory of *p-n* Junctions in Semiconductors and *p-n* Junction Transistors." The term junction as it was used in the article and as it is generally understood is the contact area between regions in a crystal having *p* and *n* characteristics. Such a junction has the properties of a rectifier.

The original transistors were not satisfactory, because it was not possible to make them by inserting wires in a crystal with any degree of uniformity, and they were noisy and mechanically weak. In Shockley's article he proposed that the transistor could be much improved by using a single crystal into the faces of which the necessary impurities were introduced. Firm connections could then be made at the required points, eliminating the unreliable wire contacts. The first such transistors were introduced commercially in 1951, but production in quantity was still not possible because silicon and germanium contained traces of unwanted impurities that interfered with satisfactory performance. These were later found to be gold in silicon and copper in germanium, in such small quantities as to defy chemical analysis.

The impurities problem was solved in 1954 when William G. Pfann, a metallurgist at Bell Laboratories, discovered a new method of refining silicon and germanium that yielded products of very high purity. To accomplish this result a long piece of the material was heated in a narrow zone by means of a high-frequency field. The molten zone was moved slowly along the length of the bar. It was found that the impurities were swept along ahead of the melted substance, leaving almost perfectly pure material behind. When the entire length of the bar had been covered in this manner, the impurities were concentrated at one end and could be removed.

With these purified materials transistors of excellent quality were produced in 1955. There were further improvements in manufacturing following one upon another, such as producing a junction in a single crystal during growth. Slices cut from such a crystal could be made into transistors. A more common method was to form junctions by putting small dots of the required impurity on opposite sides of a thin piece of crystal and baking at a temperature high enough to cause the impurities to penetrate the surface of the semiconductor material. Later in 1955 the diffusion technique was introduced in which a slice of germanium or silicon was heated in an atmosphere containing a vapor of the desired impurity, with the result that a very small amount of the impurity was deposited upon the semiconductor and penetrated into it. This method, far superior to any previously employed, was used in the production of uniform commercial transistors. What was even more important, it was possible to manufacture transistors that were capable of operating at frequencies in the hundreds of megacycles.

THE TRANSISTOR INDUSTRY

Practically all of the early development work on transistors was done at Bell Laboratories or at Western Electric Company. In 1949 the Bell Company was under pressure from the federal government to divorce the Western Electric Company, its subsidiary. Partly because of this pressure and partly for other reasons, Western Electric Company offered to license other manufacturers to produce transistors by paying a flat fee of $25,000. Later the field was thrown wide open without the payment of a fee.

This wide-open situation in the transistor industry caused a rapid growth, so that by 1954 there were twenty-five or more manufacturers in the field in the United States who were carrying on research. Although the Western Electric Company had a big lead at first, some of the competitors soon developed advanced technologies that caused the industry to progress at a tremendous rate. Among those who made significant contributions to transistor science and technology were Texas Instruments, RCA, Philco, Transitron, General Electric, Westinghouse, Hughes, Sylvania, and Raytheon.

Electrical manufacturers and research laboratories in foreign countries were soon engaged in transistor work and made contributions of immense value. Dr. Leo Esaki, the Japanese physicist, invented the tunnel diode, in which gallium arsenide was used as a semiconductor. European laboratories in Holland, Germany, England, and France were also engaged in meaningful research at an early date.

As a result of Dr. Esaki's work the General Electric Company produced a gallium arsenide transistor that was capable of frequencies of 4000 to 5000 megacycles. Experiments have been conducted with many other semiconductor materials including bismuth telluride, tungsten carbide, silicon carbide, and even diamond.

The transistor has made possible the production by the millions of portable radios. It has made available small portable transmitters, walkie-talkies, and very small tape recorders. Transistors have replaced vacuum tubes to a large extent for such varied uses as radio, television, telephone equipment, and computers. The transistor has led to the development of miniaturization by the use of integrated circuits in which entire circuits including transistors, resistors, capacitors, and their connections are etched on tiny chips. A whole new field called microelectronics has been born making possible not only a reduction in size and cost but also increased reliability and an extension of the upper frequency range.

Atomic Energy, Government Research, Nuclear Fusion

ATOMIC ENERGY

Scientists had suspected since 1898 that the atom contained large amounts of energy that in some cases could be released, and they also surmised that matter and energy were mutually convertible. It was in 1898 that the Lorentz transformations based on the Michelson-Morley experiments were announced. The equivalence of matter and energy was further emphasized by Einstein in 1905, at which time he made public his now-famous equation $E = mc^2$ Actual proof of the availability of such energy came with the experiments of Rutherford in 1919 and again in 1932 by Cockroft and Walton, when they succeeded in splitting the lithium atom. In 1932 also, N. Feather, another British scientist, succeeded in splitting the nitrogen atom by bombardment with neutrons.

Enrico Fermi, professor of physics at Rome, discovered in 1934 that when uranium atoms were bombarded with neutrons the uranium became radioactive, and in 1937 he found that the radioactivity was due to the creation of several isotopes of uranium.

Otto Hahn and Fritz Strassman of Berlin discovered in 1939 that in the Fermi experiment not only were isotopes produced, but uranium atoms were split to form pairs of other elements having atomic weights about half that of uranium. Accompanying this split there was the release of a large amount of energy.

The various isotopes of uranium were found to be fissionable in varying degrees. A slow neutron is sufficient to split a U-235 atom, whereas to split a U-238 atom requires a fast neutron. Based on the knowledge that vast amounts of energy were available from the atom, it became evident that under the proper conditions a chain reaction would be produced. In order to test these theories, it was necessary to obtain a quantity of one or more of the readily fissionable isotopes of uranium. The separation of a small quantity of U-235 was first accomplished by A. O. Nier of the University of Minnesota in 1939, using a modified Aston mass spectrograph. Experiments with this material proved beyond a doubt that such fissionable isotopes could indeed provide energy in undreamed of amounts.

Nuclear research using accelerators, or atom smashers, as they were called at the time, began to yield important results in the transmutation of elements and in the production of isotopes. As early as 1919 Rutherford produced oxygen and hydrogen from helium and nitrogen with the use of natural radiation by alpha particles from a radioactive source. In 1932 Cockcroft and Walton, who were researchers in the Cavendish Laboratory at Cambridge under Rutherford, obtained helium from hydrogen and lithium in the first accelerator experiment. Their apparatus was a puny affair compared with the many-million-volt accelerators of a few years later. Powerful accelerators were required to reach the nuclei of the heavy elements such as uranium.

Within a very short time after the completion of these accelerators, very significant results began to appear. In 1937 Perrier and Segrè discovered the element technetium, which had been produced by the bombardment of molybdenum with neutrons or deuterons.

Francium was discovered by Mlle Marguerite Perey in 1937 as a disintegration product of actinium.

Neptunium was first identified by P. H. Abelson and Edwin M. McMillan in 1939. It was produced by the bombardment of U-238 with slow neutrons and was the first of the new transuranium elements.

Astatine was found in 1940 as the result of the bombardment of bismuth with alpha particles by Corson, MacKenzie, and Segrè.

Plutonium, one of the more important new elements, was discovered in 1940 by Kennedy, McMillan, Seaborg, Segrè, and Wahl. It is formed by the decay of neptunium.

Other elements discovered during this period or a little later included americium, curium, berkelium, californium, einsteinium, fermium, and mendelevium. Nearly all of these new elements were discovered at Berkeley by researchers who in addition to those already mentioned included Glenn T. Seaborg, Joseph W. Kennedy, Ralph A. James, Leon O. Morgan, and Albert Ghiorso.

NUCLEAR RESEARCH FOR THE UNITED STATES GOVERNMENT

Hahn and Strassman of Germany split the uranium atom in 1939, the year in which World War II began. Scientists who were familiar with the possibilities presented by this discovery feared that Germany might put together an atomic bomb which would have terrible consequences for the Allied cause.

The fears were presented to President Roosevelt in a letter written by Albert Einstein on October 11, 1939, but the problem was already under attack in many laboratories throughout the United States and also in England. At the beginning of the year 1940 the only readily fissionable material was U-235 and the problem was to find methods of concentrating the 0.7 percent of U-235 that occurs naturally in uranium. By the end of the year a new fissionable element, plutonium, was discovered.

Altogether four possible methods of separating U-235 were under consideration. These were the gaseous diffusion process, begun at Columbia University; the thermal diffusion process, invented by P. H. Abelson and under test at the Naval Research Laboratory, near Washington; the centrifugal process at the University of Virginia; and the mass spectrograph, electromagnetic, or calutron process at Berkeley.

In order to coordinate these scattered efforts, the National Defense Research Committee was appointed in June 1940, under the leadership of Dr. Vannevar Bush who was succeeded a year later by Dr. James B. Conant. Progress was slow, and it was not until May 1942 that the committee met to decide which method would produce the greatest quantity of U-235 in the shortest time. The committee was unable to make a choice and decided, therefore, to build facilities using three of the available methods: gaseous diffusion, centrifuge, and calutron. Before the plant was built, however, the centrifuge method was dropped.

The committee's recommendation was accepted by the War Department and Colonel James C. Marshall of the Manhattan Engineer District was selected to head the project, which came to be known as the Manhattan Project. Within a short time Brigadier General L. R. Groves succeeded Colonel Marshall; his principal assignment was the construction of the facilities needed for the large-scale production of U-235 and later of plutonium.

A group of leading nuclear scientists was assembled at the University of Chicago where the new Metallurgical Laboratory had been completed. The primary purpose of this group was to build an atomic pile or self-sustaining reactor. This group was headed by Dr. Arthur H. Compton and included Enrico Fermi, Glenn Seaborg, Isadore Perlman, Leo Szilard, Eugene Wigner, Herbert Anderson, Walter Zinn, and Samuel Anderson. The Argonne National Laboratory, which was under construction at this time, was to house the proposed reactor, but when construction of Argonne was delayed it was decided to build a smaller one under the stands at Stagg Field. Work on this reactor was begun on November 7, 1942, and it went into service on the afternoon of December 2, 1942.

The Oak Ridge, Tennessee, plant for the separation of U-235, which included a new town for its workers, was rushed to completion. At first it contained only the gaseous diffusion and calutron separation facilities. The calutron portion was first to go into operation, and although it did produce U-235 it was found to be inefficient. The gaseous diffusion plant went into operation January 20, 1945, and it, too, was rather disappointing at first.

After it was found that production fell short of the desired quantities, it was decided that if partially enriched uranium could be fed into the production lines the process would be speeded up considerably. The decision was made, therefore, to build a thermal diffusion plant that would yield uranium enriched to at least double the natural amount of U-235. The new plant was ready in late 1944.

A plutonium plant, called the Hanford Engineering Works, was built in the State of Washington. It was begun in March 1943 and, in addition to the plant, included housing for thousands of workers. In the production of plutonium on a large scale, as in the laboratory, U-238 is bombarded with slow neutrons to form neptunium that then decays to produce plutonium. In order to carry out this process on a production basis, a very large reactor was built. After only eighteen months spent on construction, Hanford was ready. The reactor went critical a few minutes after midnight on September 27, 1944, but it had operated only a few hours when it gradually failed. Some time passed before the trouble was found, when it was discovered that the reactor was poisoned by xenon 135. Fermi finally determined that the difficulty could be overcome by operating the reactor at a higher rate. Production was resumed on Christmas Day of 1944.

LOS ALAMOS LABORATORY AND THE ATOMIC BOMB

The Los Alamos (New Mexico) Laboratory was begun in March 1943, at which time scientists from many universities in the United States were brought together under the leadership of J. R. Oppenheimer. The purpose of this group was to construct an atomic bomb. While they knew, in a general way, what was required to put together such a device, no one had any precise information as to what constituted a critical mass of uranium or how the components could be brought together in the exceedingly brief time of a few millionths of a second. These problems were solved, one by one, with the greatest of care. The first experimental bomb was exploded at Alamogordo, New Mexico, at 5:30 A.M., July 16, 1945. On August 6, 1945, an atomic bomb was dropped on Hiroshima, Japan, and three days later a second bomb obliterated Nagasaki.

ATOMIC ENERGY COMMISSION

In July 1946, about a year after the end of the war, the Congress of the United States established the Atomic Energy Commission, which since that time has had the responsibility of providing radioactive materials, not only for military purposes, but more importantly for peacetime uses. In twenty years atomic energy became competitive with coal and other fossil fuels, and there has been a tremendous expansion in medical uses of radioactive materials, opening an entirely new field in medical practice. The use of radioactive materials as tracers has found wide applications, not only in medicine, but also in plant genetics and even in industry.

NUCLEAR POWER PLANTS

Nuclear power plants are electric generating stations that use radioactive materials for fuel. They produce heat by atomic fission and generate steam for the operation of turboelectric generators. Eight basic types of reactors for use in these plants have been operated or tested experimentally. These are pressurized water, boiling water, sodium graphite, fast breeder, homogenous, organic cooled and moderated, gas cooled, and high temperature gas cooled.

Reactor assemblies, in addition to the fuel core, contain a coolant such as water, gas, liquid metals, or other materials to carry the heat from the atomic pile to a heat exchanger. In the case of a boiling water reactor, however, steam is generated directly in the reactor tank. Reactor assemblies also include a moderator, which is most commonly water or graphite, and finally there must be provision for control, which is usually accomplished by means of control rods.

The moderator is designed to change the fast neutrons, with velocities averaging 10,000 miles per second, to slow (or thermal) neutrons, having velocities of about 1 mile per second. The slow neutrons are much more effective in splitting U-235 atoms than the fast neutrons.

Control rods are necessary to hold the rate of fission at

the level required for power demands or to shut down the reactor. There are other methods of controlling the production of heat but control rods are the most commonly used. They may be composed of boron steel, boron carbide, gadolinium, samarium, or other materials.

In the United States the fuel used in nuclear reactors is most often enriched uranium containing from 1.5 to 5 percent U-235. In Britain natural uranium metal is used in most cases because the facilities for the production of U-235 are not sufficient for the purpose.

When a U-235 atom is split, about 2.5 neutrons are released on the average, but some of these are lost. At least one of the neutrons must be available for splitting another atom if the chain reaction is to be maintained.

Most of the nuclear power generating stations in the United States are of the pressurized-water or boiling water type. In England, which at present has most of the world's nuclear generating capacity, gas-cooled reactors have been used exclusively. The gas employed has been mostly carbon dioxide, but with higher temperatures in the newer plants it has become necessary to change to helium.

The first large-scale nuclear power plant in the world was Calder Hall in England, which began operating in October 1956. In the United States the first nuclear power plant designed solely for power generation was the Shippingport station in Pennsylvania, which began the production of electricity on December 2, 1957. Calder Hall was a gas-cooled plant using carbon dioxide and Shippingport was a pressurized-water plant.

At the end of 1966, England had 4,048,000 kilowatts of nuclear generating capacity in operation, and the United States had 1,956,000 kilowatts. At that time, new capacity under construction or planned was 5,290,000 kilowatts in

England and 9,285,000 kilowatts in the United States. At the end of the year 1969, the United States had 5,235,000 kilowatts of nuclear capacity in operation and had 65,365,000 kilowatts under construction or planned, with the latest completion date set at 1977 but with most of the capacity to be ready by 1974.

NUCLEAR FUSION

Although the energy liberated in nuclear fission is very great, the energy available from nuclear fusion is even greater but much more difficult to achieve. Fission occurs in the heavier elements, whereas fusion is possible only in the lighter elements, notably hydrogen and tritium (the triple mass isotope of hydrogen). Helium apparently cannot be used in the fusion process, but it is an end product of fusion. It is possible that lithium may be suitable for this purpose. The great difficulty in achieving fusion is that it requires temperatures in the millions of degrees and very high pressures. Such conditions prevail in the interior of our sun and in the stars, but so far a controlled fusion process has been impossible. Hydrogen bombs have been exploded with the use of an atomic bomb to trigger the reaction. Much work has been done in seeking a method of using the fusion process for the production of useful energy, but it is evident that no substance on earth could contain a fusion reaction at the required temperatures and pressures. So far the most promising method is one in which the required conditions are obtained by confining the process within a powerful magnetic field (pinch effect), but the solution seems a long way off.

WHITHER

When Joseph Priestley, back in the eighteenth century, wrote his book, *The History and Present State of Electric-*

ity, he predicted that new developments in electrical science would far outstrip anything yet conceived, but some of his contemporaries were sure that nothing more could be learned about the subject. Considering the advanced state of electrical science today, we might be forgiven if we believed that man had reached the limit of his capabilities in this direction, but we must recognize that each new step has uncovered more questions than it has provided answers. Therefore, if the past is in any way a guide to the future, we may be sure that still greater achievements lie ahead. As we view the accomplishments of the last half century we are amazed at the ability of men of science to learn of things which the "eye hath not seen nor ear heard." At the same time every truly great scientist must view with awe, wonder, and admiration the great works of creation, whose mysteries he has explored.

BIBLIOGRAPHY

1

Davis, Sir J. F. *The Chinese*. 2 vols. New York: W. W. Norton Co., 1845-1848.

Dibner, Bern. *Ten Founding Fathers of the Electrical Science*. Norwalk, Conn.: Burndy Library, 1954.

Du Halde, Jean Baptiste. *Déscription de la Chine*. Paris: P. G. Mercier, 1735.

Gilbert, William. *De Magnete*. Translation by Paul Fleury Mottelay. New York: Dover Publications, 1958.

Humboldt, Friedrich. *Cosmos*. Translation by E. C. Otté and Others. London: Bohn's Scientific Library, 1871.

Munro, John. *Pioneers in Electricity*. London: Religious Tract Society, 1890.

Pliny, the Elder. *Natural History*. 6 vols. Translated by John Bostock and H. T. Riley. London and New York: George Bolt and Sons, 1893.

Routledge, Robert. *A Popular History of Science, Early Discoveries*. London and New York: G. Routledge and Sons, 1881.

Simons, Eric N. *Metals, History*. London: Dobson, 1967.

2

Canby, Courtlandt, ed. *History of Electricity*. New York: Hawthorn Books, 1963.

Dibner, Bern. *Early Electrical Machines*. Norwalk, Conn.: Burndy Library, 1957.

Fleming, Arthur, and Brocklehurst, H. J. S. *A History of Engineering*. London: Black, 1925.

Hulme, E. M. *Renaissance and Reformation*. New York: The Century Co., 1915.

Loeb, L. B. *Fundamentals of Electricity and Magnetism*. 2nd ed. New York: John Wiley and Sons, 1938.

Parsons, William B. *Engineers and Engineering in the Renaissance*. Cambridge, Mass.: The M.I.T. Press, 1968.

Priestley, Joseph. *The History and Present State of Electricity*. 1st ed. London: J. Dodsley, 1767. 2nd ed. London: Bathurst, 1775.

Williams, H. S. *History of Science*. New York and London: Harper and Brothers, 1910.

3

Cohen, I. Bernard. *Benjamin Franklin's Experiments*. Cambridge, Mass.: Harvard University Press, 1941.

Dibner, Bern. *Galvani, Volta*. Norwalk, Conn.: Burndy Library, 1952.

Dibner, Bern. *Oersted*. Norwalk, Conn.: Burndy Library, 1961.

Dunsheath, Percy. *Giants of Electricity*. New York: Crowell, 1967.

Mottelay, Paul Fleury. *Bibliographical History of Electricity and Magnetism*. London: C. Griffin and Co., 1922.

Still, Alfred. *Soul of Amber*. New York: Murray Hill Books, 1944.

Youmans, W. J. *Pioneers of Science in America*. New York: Appleton, 1896.

4

Dibner, Bern. *Faraday*. Norwalk, Conn.: Burndy Library, 1949.

Coulson, Thomas. "Joseph Henry." *Colliers Encyclopedia*. New York, 1967.

Faraday, Michael. *Experimental Researches in Electricity*. London: R. and J. E. Taylor, 1839.

Iles, George. *Invention and Discovery*. New York: Doubleday Page and Co., 1906.

Thompson, Silvanus P. *Michael Faraday, His Life and Work*. New York: Macmillan Co., 1898.

Welling, J. C. *Life and Character of Joseph Henry*. Cambridge, Mass.: The Riverside Press, 1904.

Youmans, W. J. *Pioneers of Science in America*. New York: Appleton, 1896.

5

Beck, W. *Die Elektrizität und Ihre Technik*. Leipzig: Ernst Wiest, 1896.

Brackett, C. F., and Others. *Electricity in Daily Life*. New York: Charles Scribner's Sons, 1891.

Kuhlman, John. *Design of Electrical Apparatus*. 3rd ed. New York: John Wiley and Sons, 1950.

Sheldon, Samuel, and Haussman, Erich. *Dynamo Electric Machinery*. New York: D. Van Nostrand Co., 1911.

6

Austin, Frank E. *Examples in Battery Engineering*. Hanover, N. H.: Published by the author, 1917.

Cooper, W. R. *Primary Batteries and Their Theory, Construction and Use*. London: The Electrician Printing and Publishing Co., 1916.

Lincoln, E. S. *Primary and Storage Batteries*. New York: Essential Books, 1945.

Schure, Alexander. *Electrostatics*. New York: Rider Publications, 1958.

Smythe, W. R. *Static and Dynamic Electricity*. New York: McGraw-Hill, 1939.

White, H. G. *Electric Bells, Alarms and Signalling Systems*. London: Rentell, n. d.

7

Bell, Eric Temple. "Gauss, the Prince of Mathematicians." In *The World of Mathematics*, edited by James R. Newman. New York: Simon and Schuster, 1956.

Kinnard, Isaac F. *Applied Electrical Measurements*. New York: John Wiley and Sons, 1956.

Lincoln, Edwin S. *Electrical Measuring Instruments*. New York: Essential Books, 1945.

Page, C. H. *U.S.A. and IEEE Standard Letter Symbols for Units*. New York: Institute of Electrical and Electronic Engineers, 1968.

Silsbee, Francis B. *Systems of Electrical Units*. Monograph 56. Washington: National Bureau of Standards, 1963.

Smith, Arthur W. *Electrical Measurements*. New York: McGraw-Hill, 1924.

8

Mabee, Carleton. *The American Leonardo: The Life of Samuel F. B. Morse*. New York: Octagon Books, 1969.

Morgan, A. P. *The Pageant of Electricity*. New York: Appleton-Century, 1939.

Prime, Samuel I. *Life of Samuel F. B. Morse*. New York: D. Appleton Co., 1875.

Youmans, W. J. *Pioneers of Science in America*. New York: Appleton, 1896.

9

Brackett, C. F., and Others. *Electricity in Daily Life.* New York: Charles Scribner's Sons, 1891.

Dibner, Bern. *The Atlantic Cable*. Norwalk, Conn.: Burndy Library, 1959.

Iles, George. *Invention and Discovery*. New York: Doubleday Page and Co., 1906.

10

Carty, John J. *Annual Report of the Smithsonian Institution*. Washington: Smithsonian Institution, 1922.

Mackenzie, Catherine. *Alexander Graham Bell*. Boston and New York: Houghton Mifflin Co., 1928.

Prescott, George B. *The Speaking Telephone*. New York: D. Appleton Co., 1879.

11

Crowther, J. G. *Famous American Men of Science*. New York: W. W. Norton Co., 1937.

Dyer, F. L., and Martin, T. C. *Edison, His Life and Inventions*. New York: Harpers, 1929.

Hammond, John Winthrop. *Men and Volts*. New York: J. B. Lippincott Co., 1941.

Howell, John W., and Schroeder, Henry. *The History of the Incandescent Lamp.* Schenectady, N. Y.: The Maqua Co., 1927.

Josephson, Matthew. *Edison*. New York: McGraw-Hill, 1959.

Prescott, George B. *The Electric Light*. New York: D. Appleton Co., 1879.

12

Armagnat, H. *The Theory, Design and Construction of Induction Coils*. New York: McGraw Publishing Co., 1908.

Barham, G. Basil. *Development of the Incandescent Lamp.* London: Scott Greenwood and Son, 1912.

Fleming, J. A. *The Alternate Current Transformer*. 3 vols. London: The Electrician Printing and Publishing Co., 1893-1894.

Lamphier, R. C. *Electric Meter History and Progress*. Springfield, Ill.: Sangamo Electric Co., 1925.

Reed, E. G. *Essentials of Transformer Practice*. New York: D. Van Nostrand Co., 1927.

13

Hammond, John Winthrop. *Men and Volts*. New York: J. B. Lippincott Co., 1941.

Houston, Edwin J. *Electricity in Everyday Life*. Vols. 2 and 3. New York: P. F. Collier and Son, 1904-1905.

Norris, Henry H. *Electric Railway Practice in 1925*. New York: American Railway Association, 1926.

14

Dunlap, Orrin E., Jr. *Radio and Television Almanac*. New York: Harper, 1951.

F.C.C. Bulletins. 2-B, 3-G, 11-S. Washington: U. S. Government Printing Office, 1966 and 1968.

F.C.C. Rules. Updated annually or oftener, Vols. 2 and 3. Washington: U. S. Government Printing Office, 1970.

Jones, Charles R. *Facsimile*. New York: Murray Hill Books, 1949.

Radio Corporation of America. *The Birth of an Industry*. New York: RCA, n. d.

Scheraga, M. G., and Roche, J. J. *Video Handbook*. Montclair, N. J.: Boyce, 1949.

15

Anderson, David L. *The Discovery of the Electron*. Princeton: D. Van Nostrand Co., 1964.

Asimov, Isaac. *Understanding Physics*. Vol. 3, *The Electron, Proton, and Neutron*. London: Allen and Unwin, 1966.

Bennett, Allen; Heikes, Robert; Klemens, Paul; and Maradudin, Alexii. *Electrons on the Move*. New York: Walker and Co., 1964.

Brown, Thomas B. *Foundations of Modern Physics*. New York: John Wiley and Sons, 1940.

Buckley, H. *A Short History of Physics*. London: Methuen, 1927.

D'Albe, E. E. Fournier. *The Life of Sir William Crookes*. London: Unwin, 1923.

Dunlap, H. A., and Tuch, H. N. *Atoms at Your Service*. New York: Harper, 1957.

Einstein, Albert. *Relativity: The Special and General Theory*. London: Methuen, 1954.

French, Anthony P. *Principles of Modern Physics*. New York: John Wiley and Sons, 1958.

Hawley, Glessner G. *The Story of the Electron Microscope*. New York: Alfred A. Knopf, 1945.

Hoffman, M. *Readings in the Atomic Age*. New York: H. W. Wilson Co., 1950.

Livingston, M. Stanley, and Blewett, J. P. *Particle Accelerators*. New York: McGraw-Hill, 1962.

Miller, D. C. *Sparks, Lightning and Cosmic Rays*. New York: The Macmillan Co., 1939.

Neal, Richard B. *The Two-Mile Linear Electron Accelerator*. From Proceedings of 75th Anniversary Symposium at Georgia Institute of Technology, February 1963.

Prime, Samuel I. *Life of Sir William Crookes*. New York: D. Appleton and Co., 1875.

Rusk, Rogers D. *Introduction to Atomic and Nuclear Physics*. New York: Appleton, 1958.

Rutherford, Ernest. *Radioactivity*. Cambridge: Cambridge University Press, 1905.

Stanford Accelerator Staff. *An Informal History of SLAC* (Stanford Linear Accelerator Center). Palo Alto: Stanford University, 1968.

White, Harvey E. *Modern College Physics*. 3rd ed. New York: D. Van Nostrand Co., 1958.

16

American Telephone and Telegraph Company. *Coaxial Cable, A Modern Communications Medium*. New York: A.T. and T. Co., 1965.

American Telephone and Telegraph Company. News Release 8-10-65. New York: A.T. and T. Co., 1965.

American Telephone and Telegraph Company. *Radio Relay, Communications by Microwave*. New York: A.T. and T. Co., 1966.

Andrews, Alan. *A.B.C.'s of Radar*. Indianapolis: Indianapolis Sams, 1966.

Booth, Andrew, and Booth, Kathleen. *Automatic Digital Computers*. Washington: Butterworth, 1965.

Bukstein, Edward J. *Digital Counters and Computers*. New York: Holt, Rinehart and Winston, 1960.

Dunlap, Orrin E., Jr. *Radar: What Radar Is and How It Works*. New York: Harper, 1946.

Hartley, Michael G. *Introduction to Electronic Analog Computers*. New York: John Wiley and Sons, 1962.

International Business Machines Corporation. *Highlights of IBM History*. Armonk, N. Y.: IBM, 1968.

National Bureau of Standards. "Computer Sciences and Technology at N.B.S." *Technical News Bulletin*. Washington: U. S. Government Printing Office, August 1967.

Sperry Rand Corporation. *A History of Sperry Rand Corporation*. 3rd ed. New York: Sperry Rand Corporation, 1964.

Thomas, Shirley. *Computers: Their History, Present Applications and Their Future*. New York: Holt, Rinehart and Winston, 1965.

17

American Telephone and Telegraph Company. *Laser, The New Light*. New York: A.T. and T. Co., 1965.

American Telephone and Telegraph Company. *The Transistor Age*. New York: A. T. and T. Co., 1965.

Chandrasekhar, S. *Plasma Physics*. Chicago: University of Chicago Press, 1960.

Delcroix, J. L. *Plasma Physics*. London and New York: John Wiley and Sons, 1965.

Marshall, Samuel L. *Laser Technology and Applications*. New York: McGraw-Hill, 1968.

Schawlow, A. L. *Masers* (Written for Colliers Encyclopedia). New York: Bell Telephone Laboratories, 1962.

"Semiconductors." *Business Week*. March 26, 1960. Pages 74-121.

18

Green, Alex E. S. *Nuclear Physics*. New York: McGraw-Hill, 1955.

Groueff, Stéphane. *Manhattan Project*. Boston: Little, Brown and Co., 1967.

Rusk, Rogers D. *Introduction to Atomic and Nuclear Physics*. 2nd ed. New York: Appleton, 1958.

Wills, J. George. *Nuclear Power Plant Technology*. New York: John Wiley and Sons, 1967.

INDEX